BLACK POWDER \\ RED EARTH: YEMEN

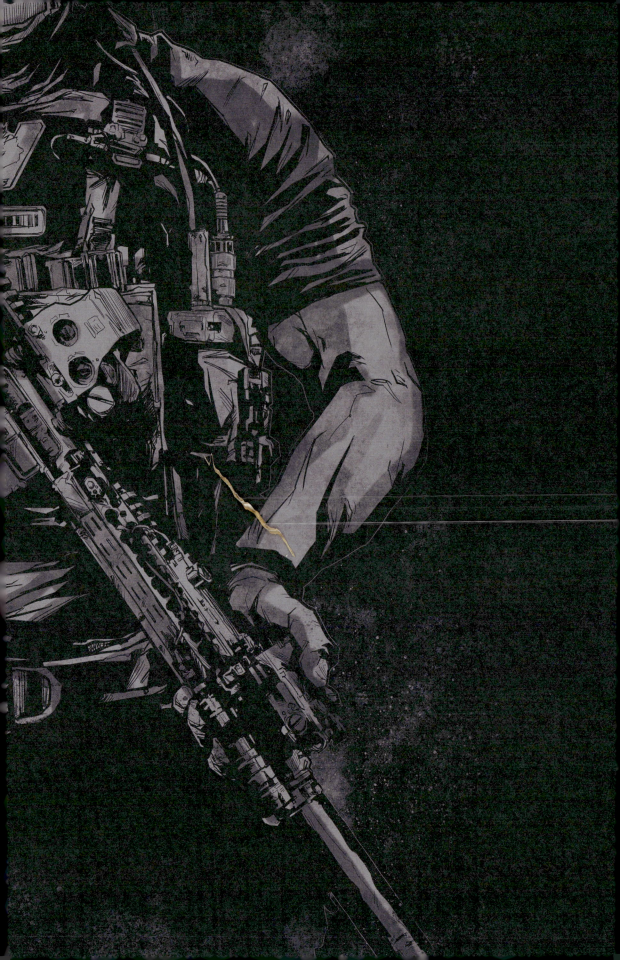

IN THE DECADE FOLLOWING THE INVASION OF IRAQ, PRIVATE MILITARY COMPANIES BECOME THE PREFERRED INSTRUMENT OF FOREIGN POLICY BY PETRO-MONARCHIES IN THE FAILED STATES OF THE MIDDLE EAST.

STAFFED WITH FORMER SPECIAL OPERATIONS COMBAT VETERANS, PRIVATE MILITARY CONTRACTORS SELECT, TRAIN AND LEAD ARMIES OF LOCAL NATIONALS IN CONVENTIONAL AND UNCONVENTIONAL WARFARE OPERATIONS.

COMBAT ACTIONS CONDUCTED BY THE PMC ARE NOT CONSIDERED ACTS OF WAR...

THEY ARE MATTERS OF FOREIGN INTERNAL DEFENSE.

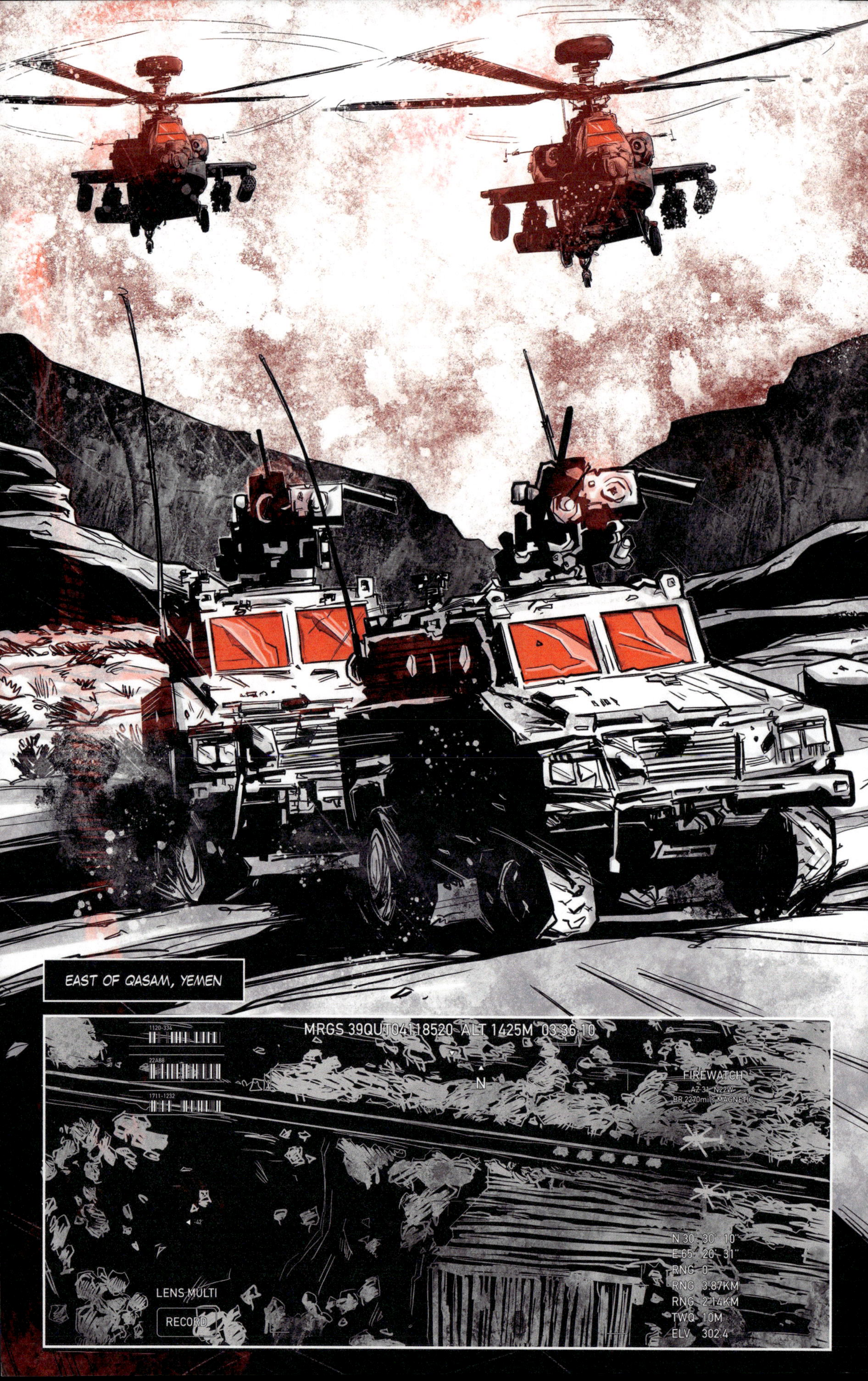
EAST OF QASAM, YEMEN
MRGS 39QUT04118520 ALT 1425M 03:36:10
1120-334
22A88
1711-1232
N
FIREWATCH
AZ 31 N22W
BR 2270mils MAGNETIC
N 30° 30' 10"
E 65° 20' 31"
RNG 0
RNG 3.87KM
RNG 2.14KM
TWQ 10M
ELV 302.4
LENS MULTI
RECORD

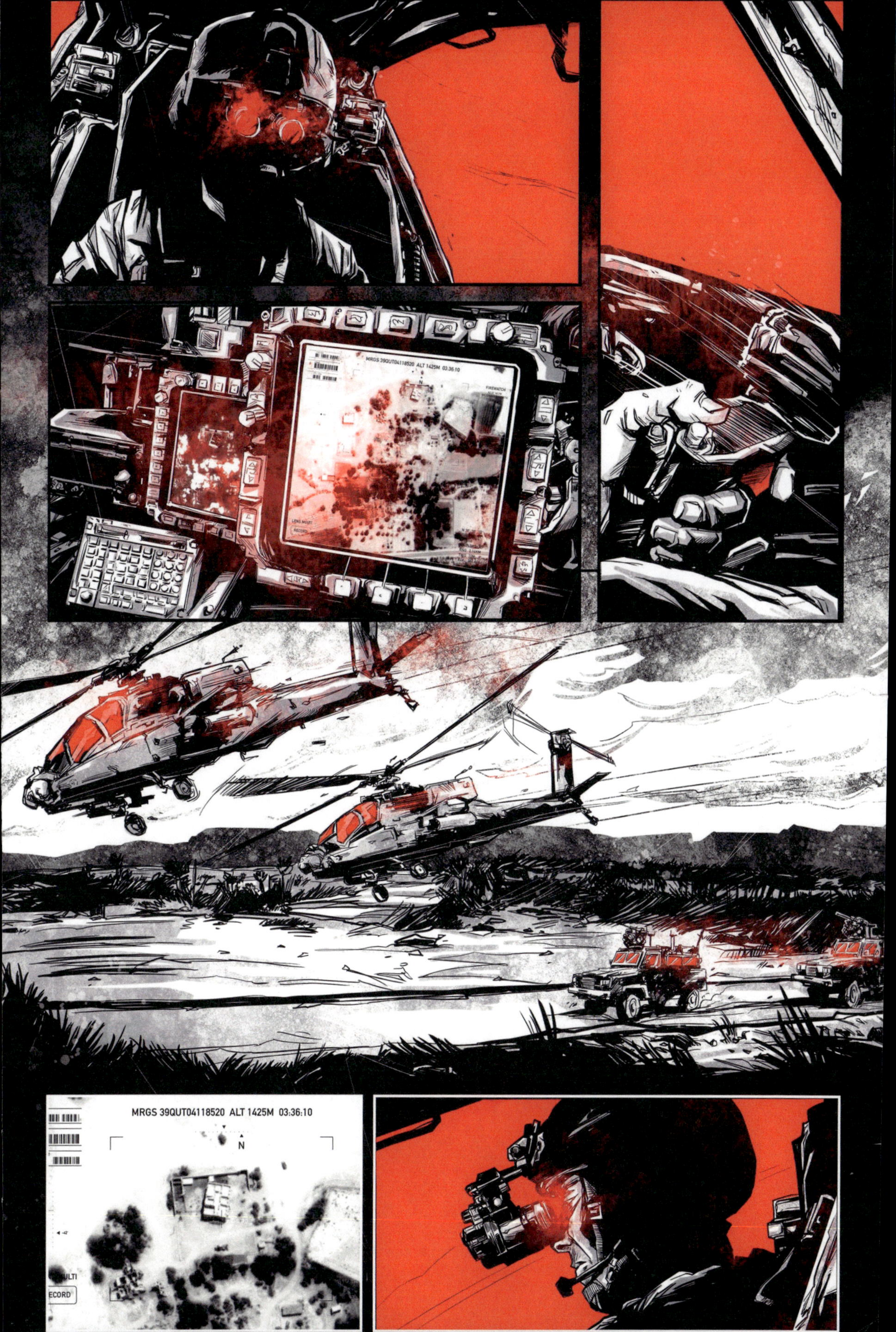
MRGS 39QUT04118520 ALT 1425M 03:36:10
N

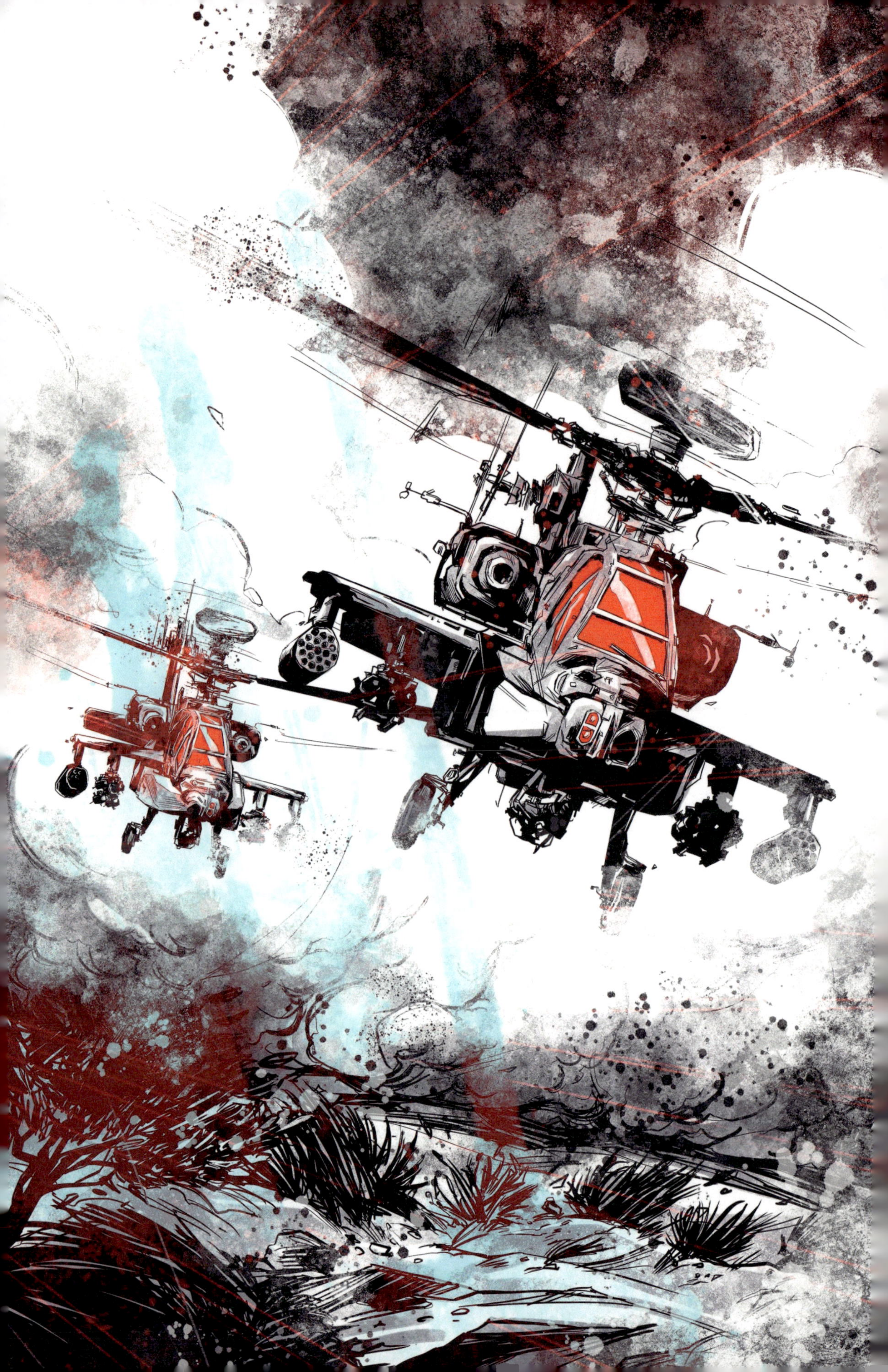

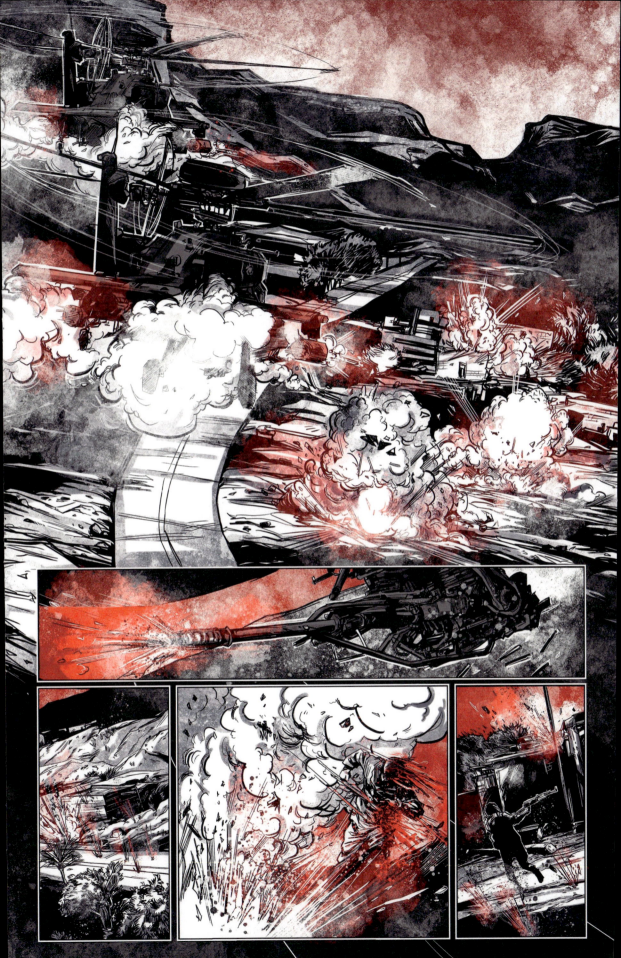

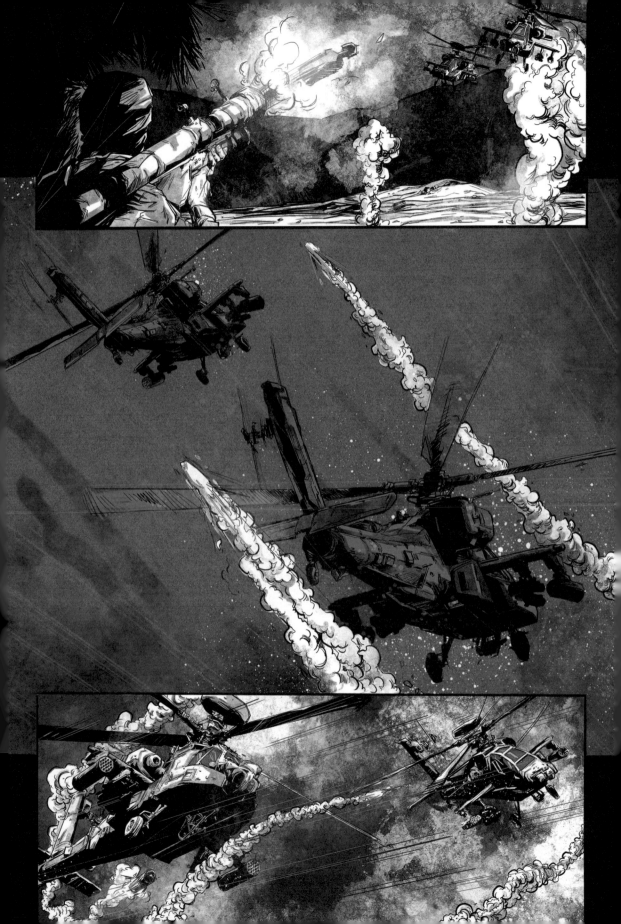

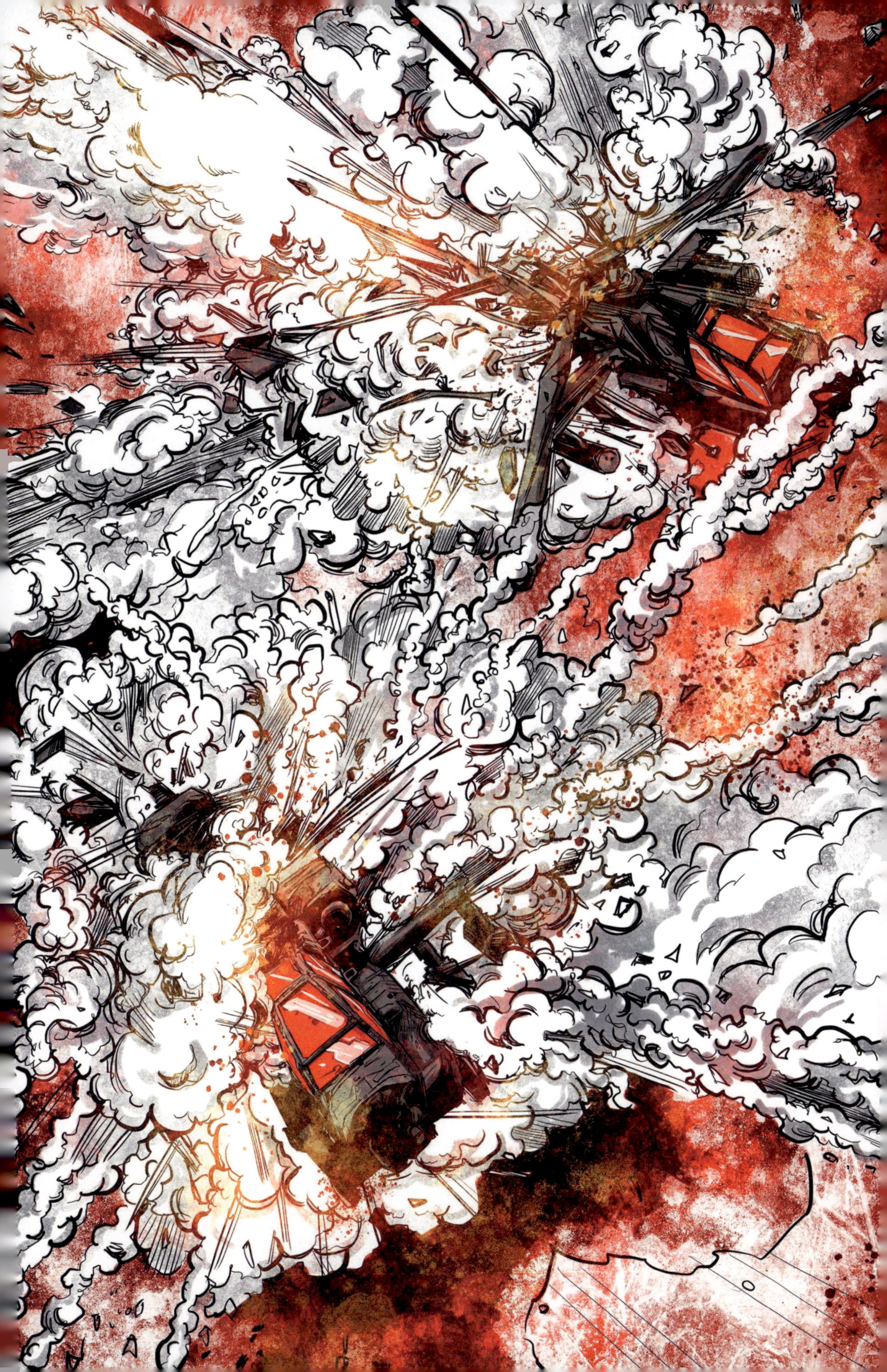

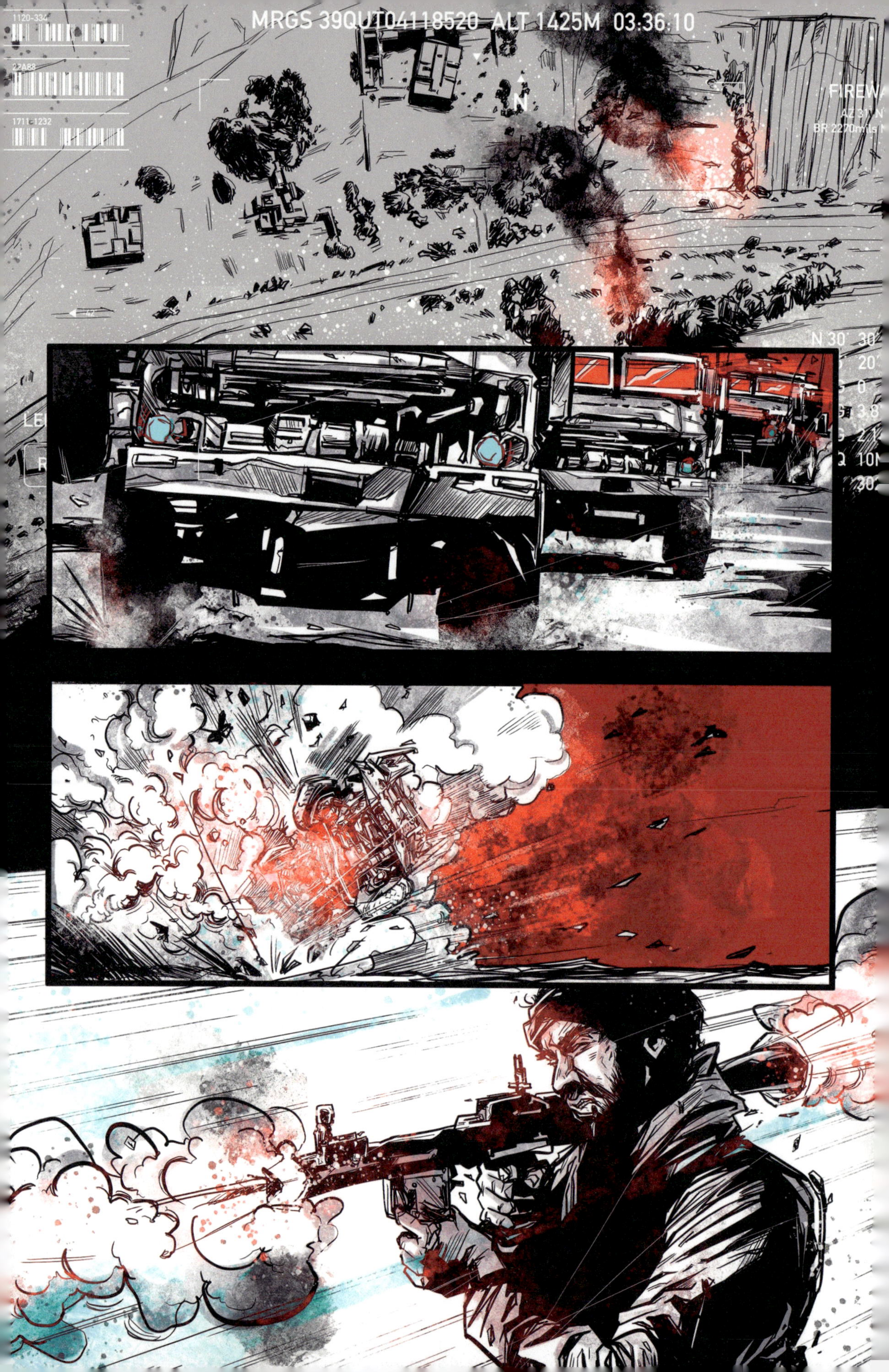
MRGS 39QUT04118520 ALT 1425M 03:36:10
1120-334
22A88
1711-1232
N
N 30

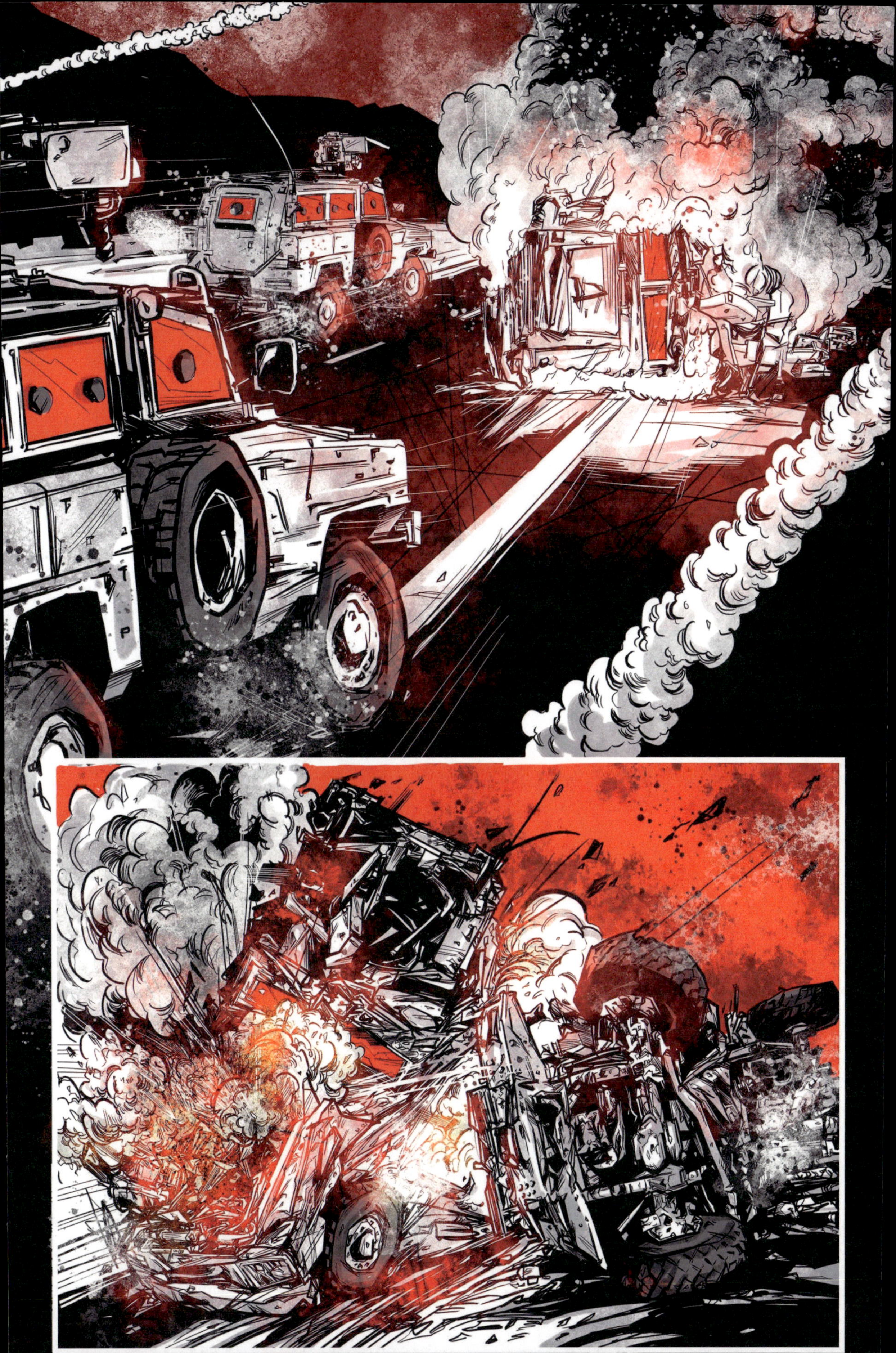

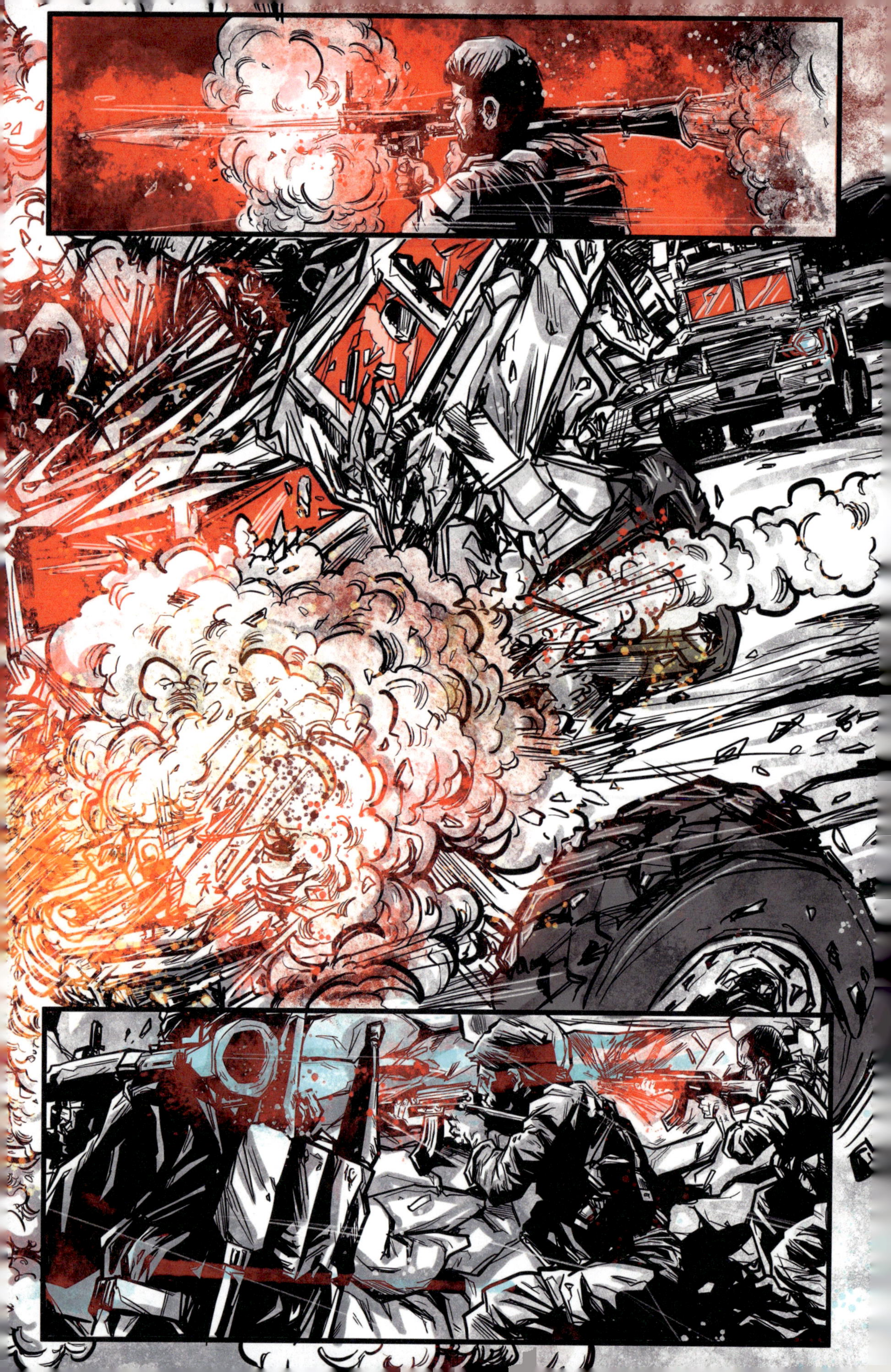

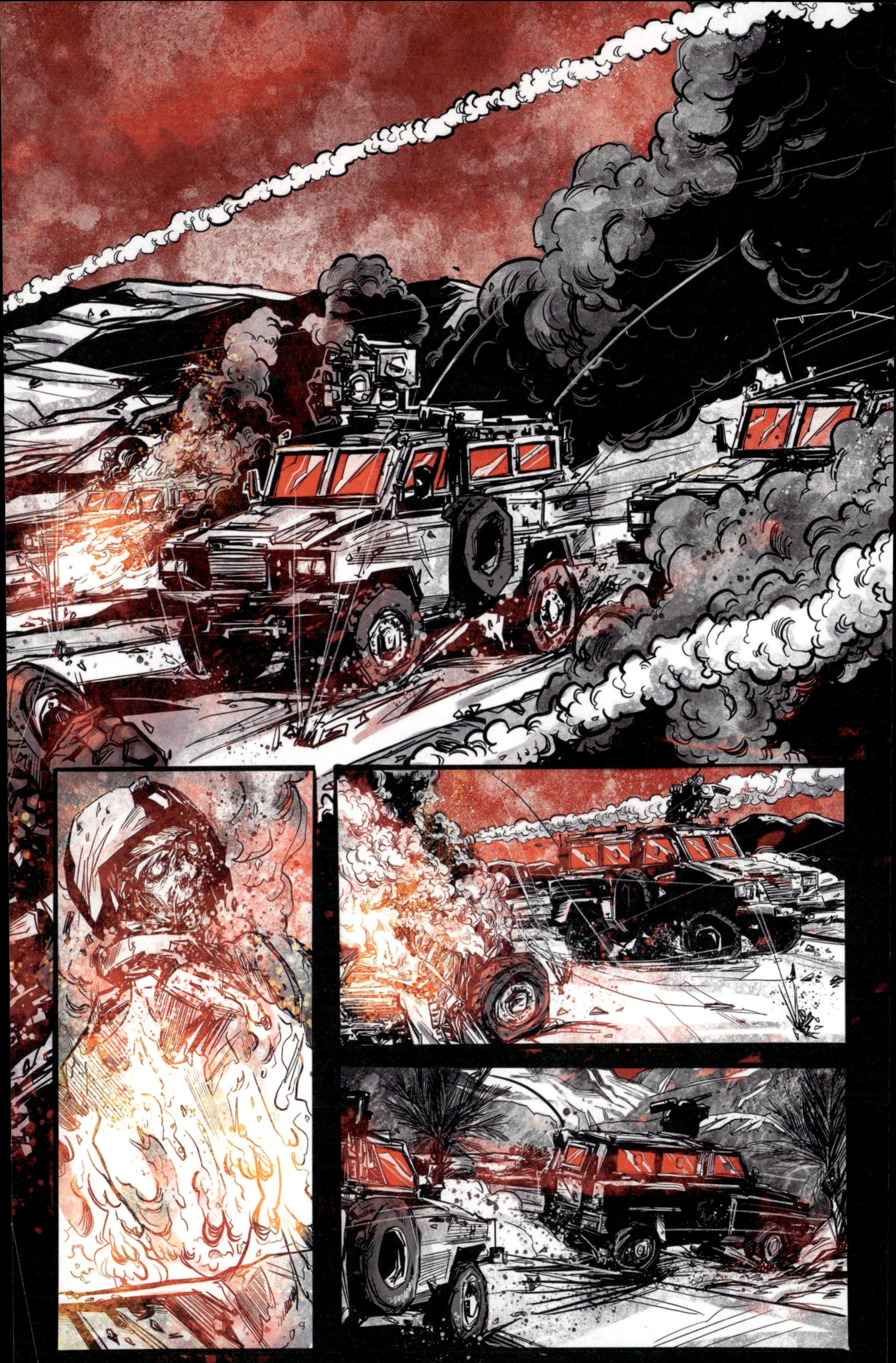

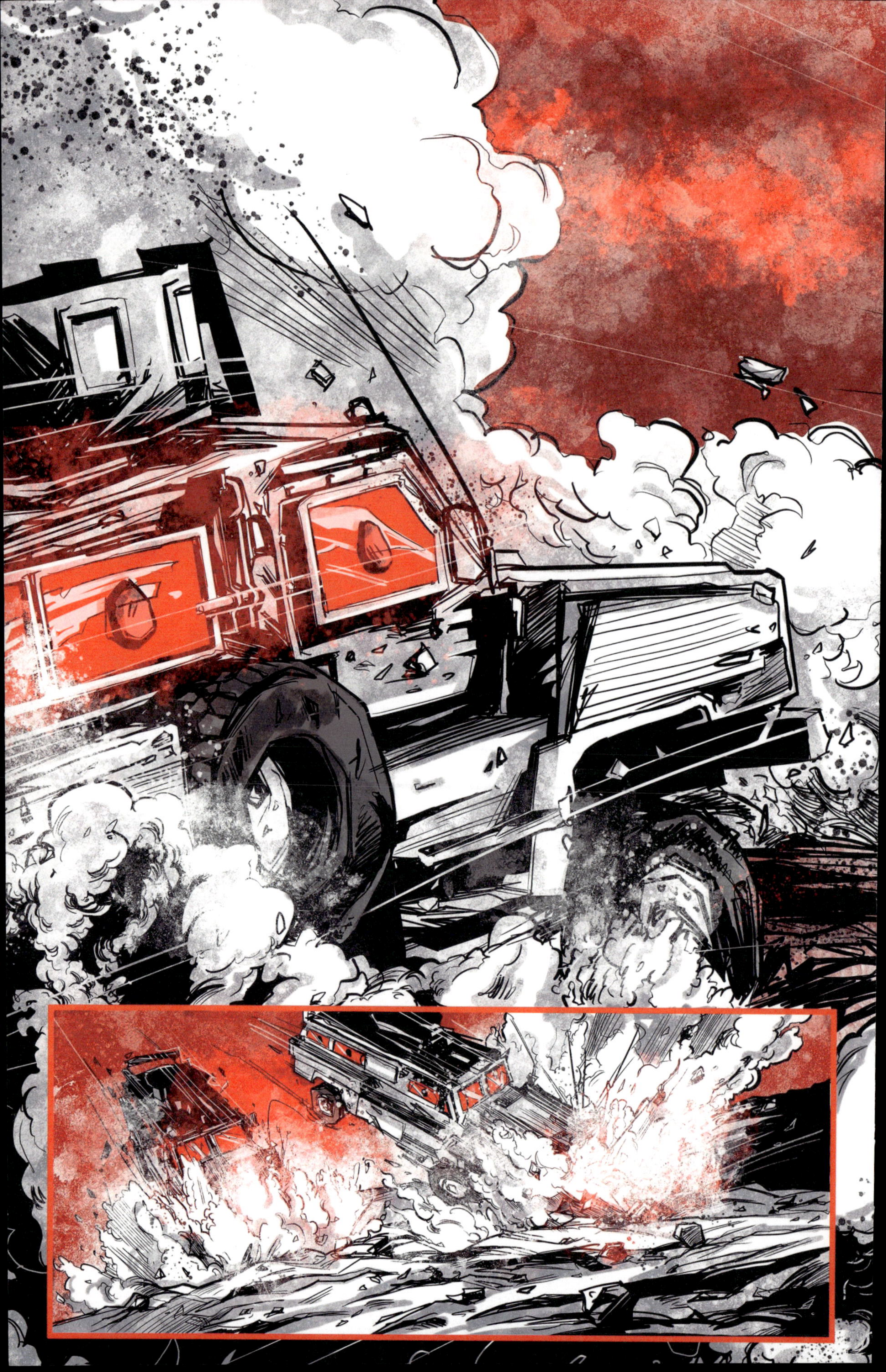

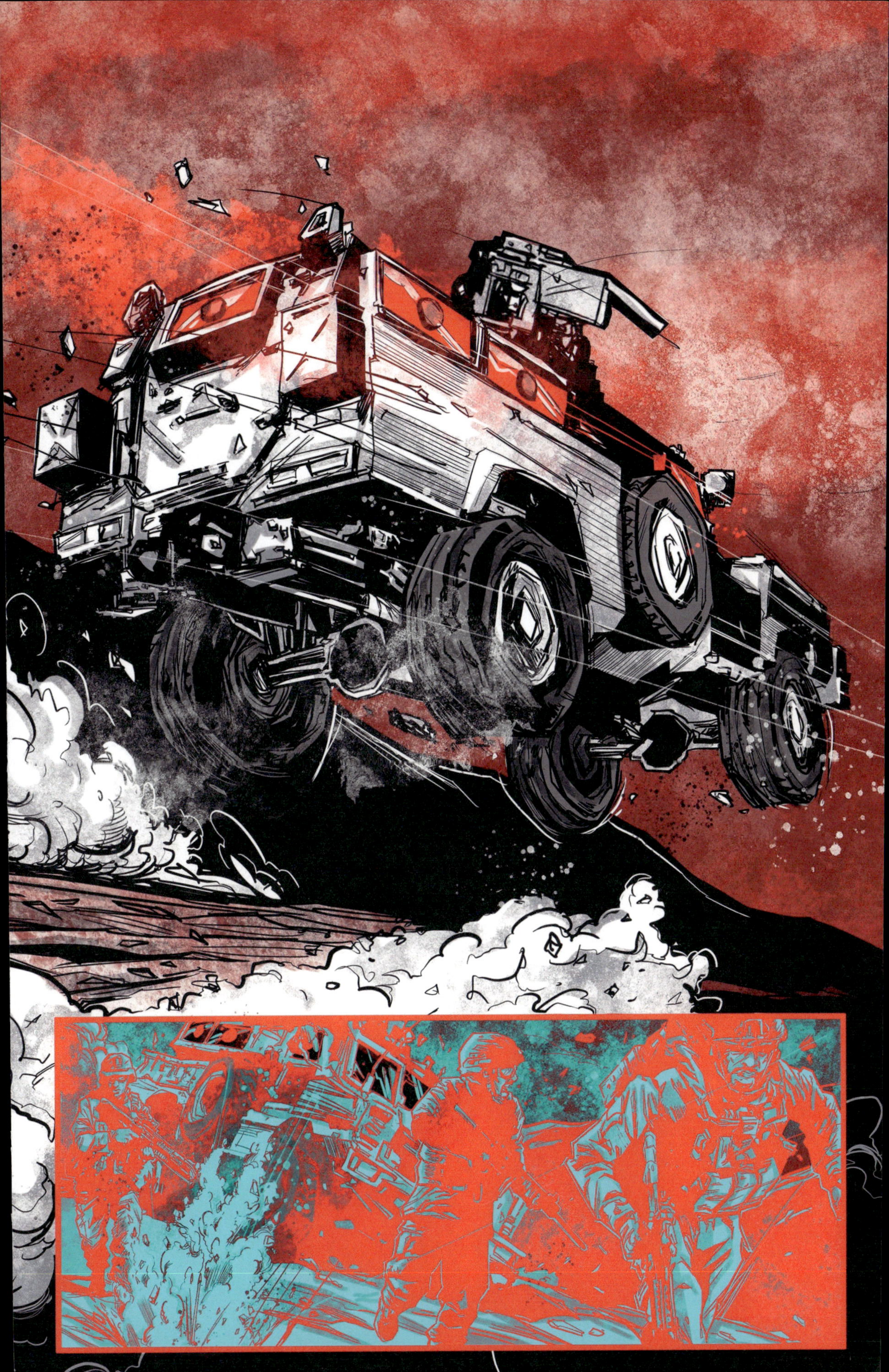

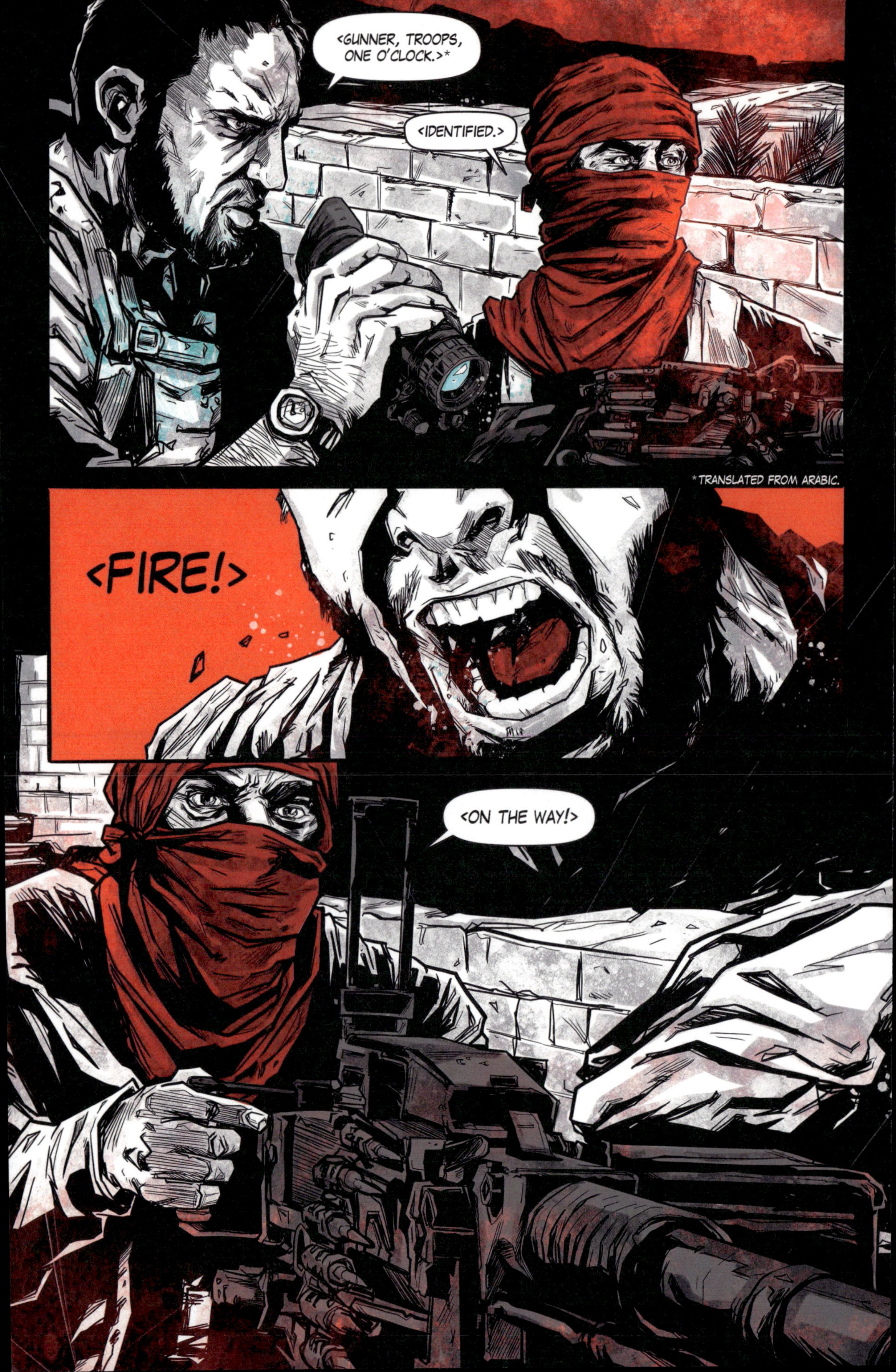
<GUNNER, TROOPS, ONE O'CLOCK.>*
<IDENTIFIED.>
*TRANSLATED FROM ARABIC.
<FIRE!>
<ON THE WAY!>

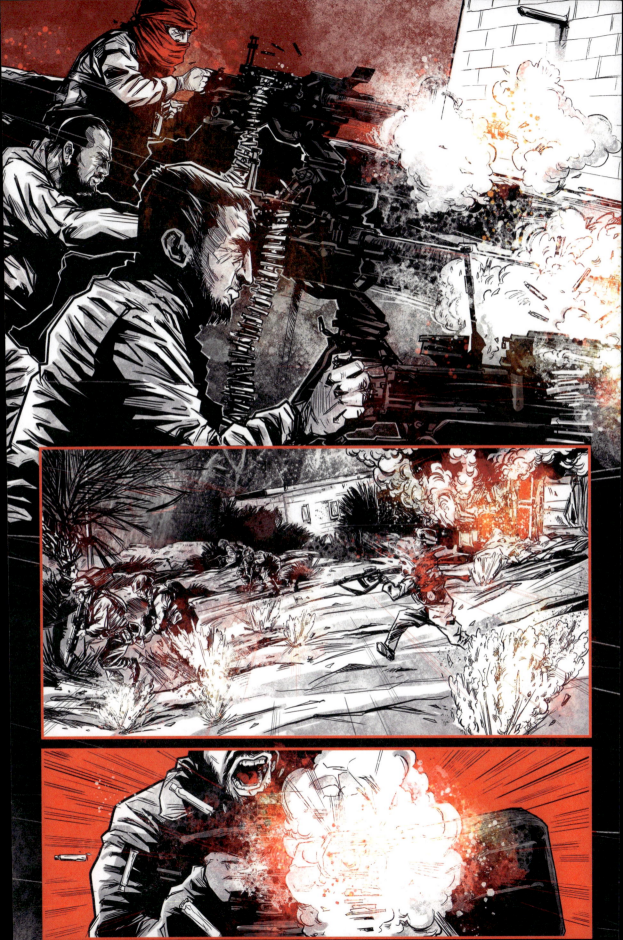

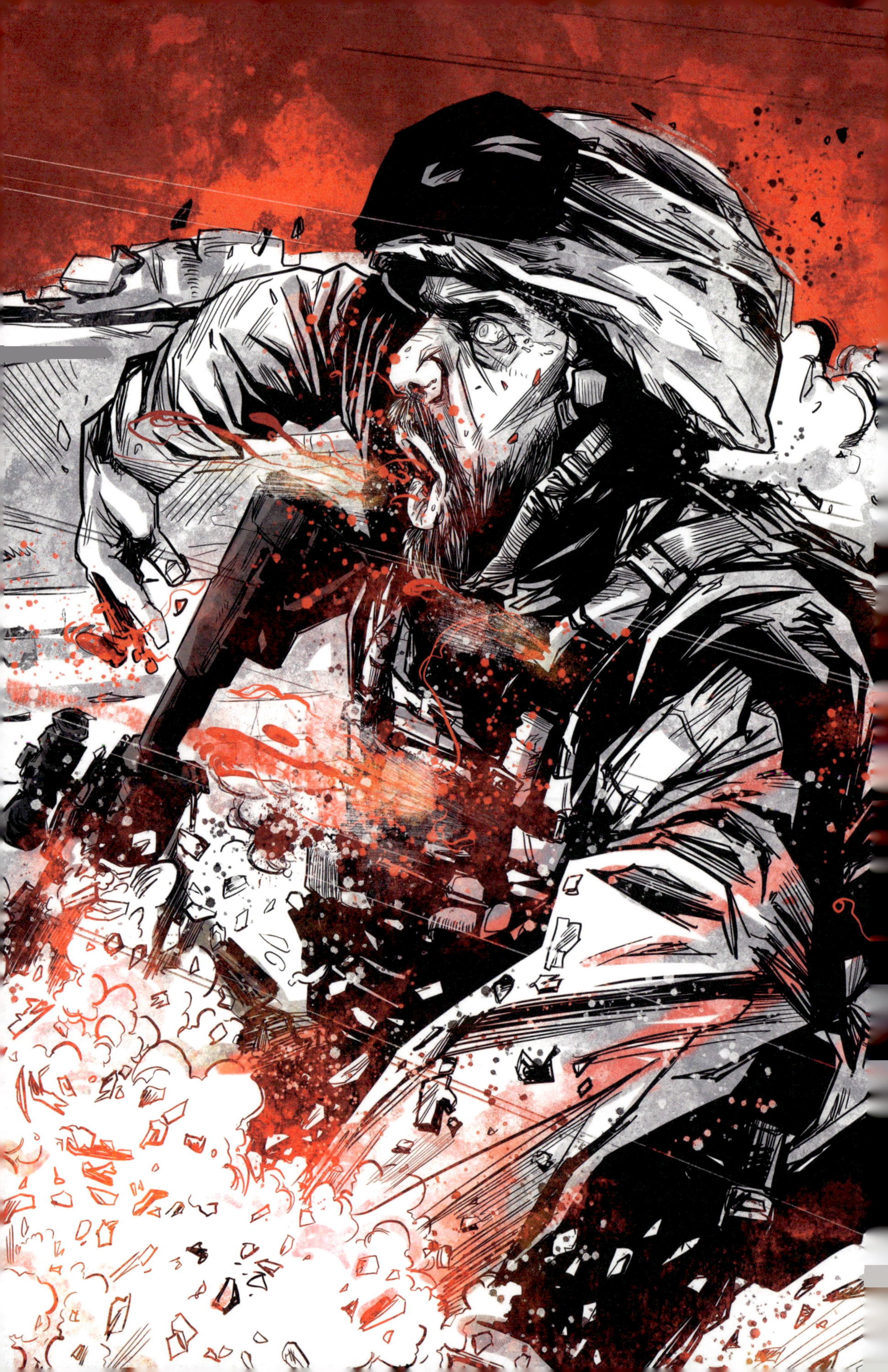

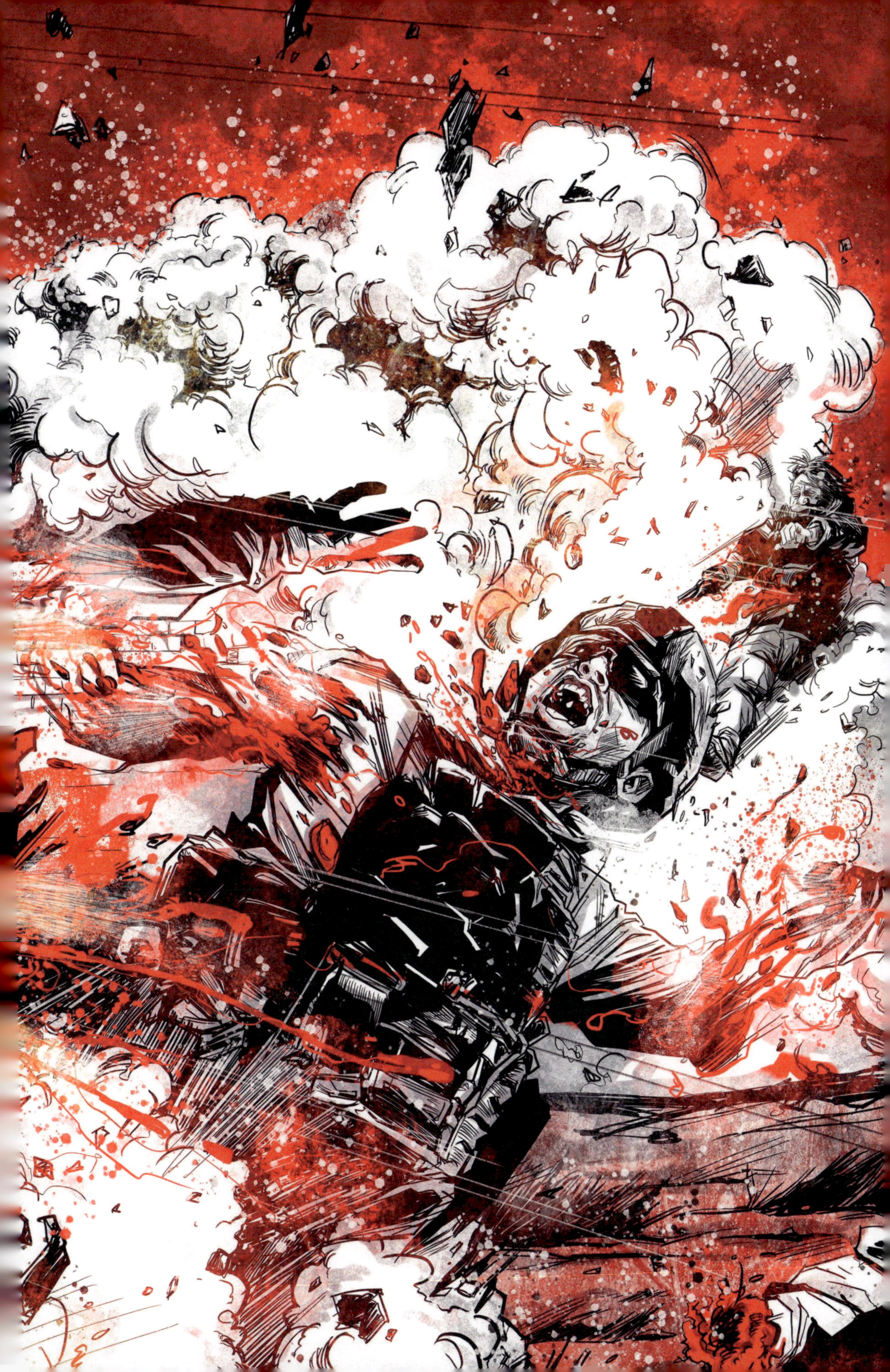

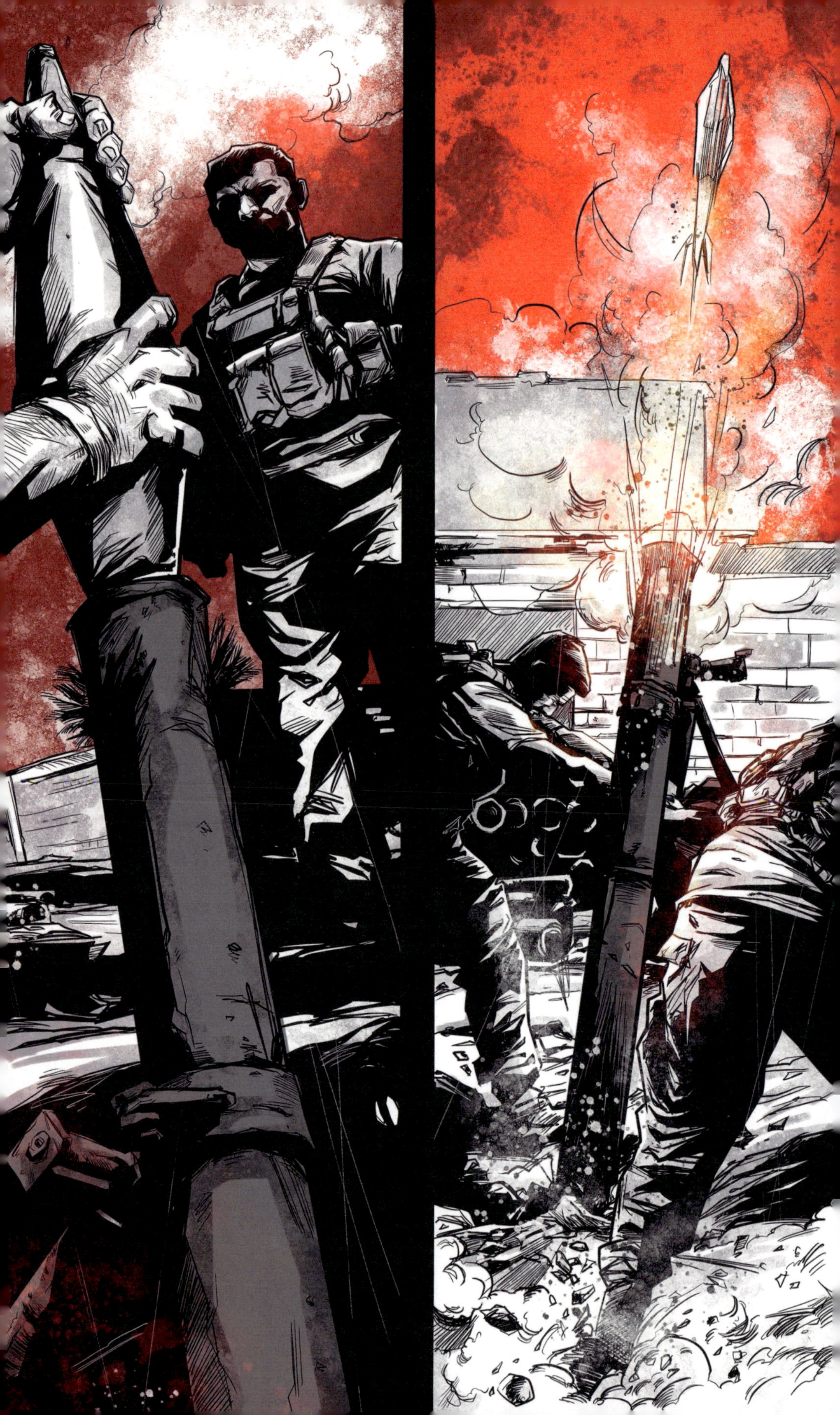

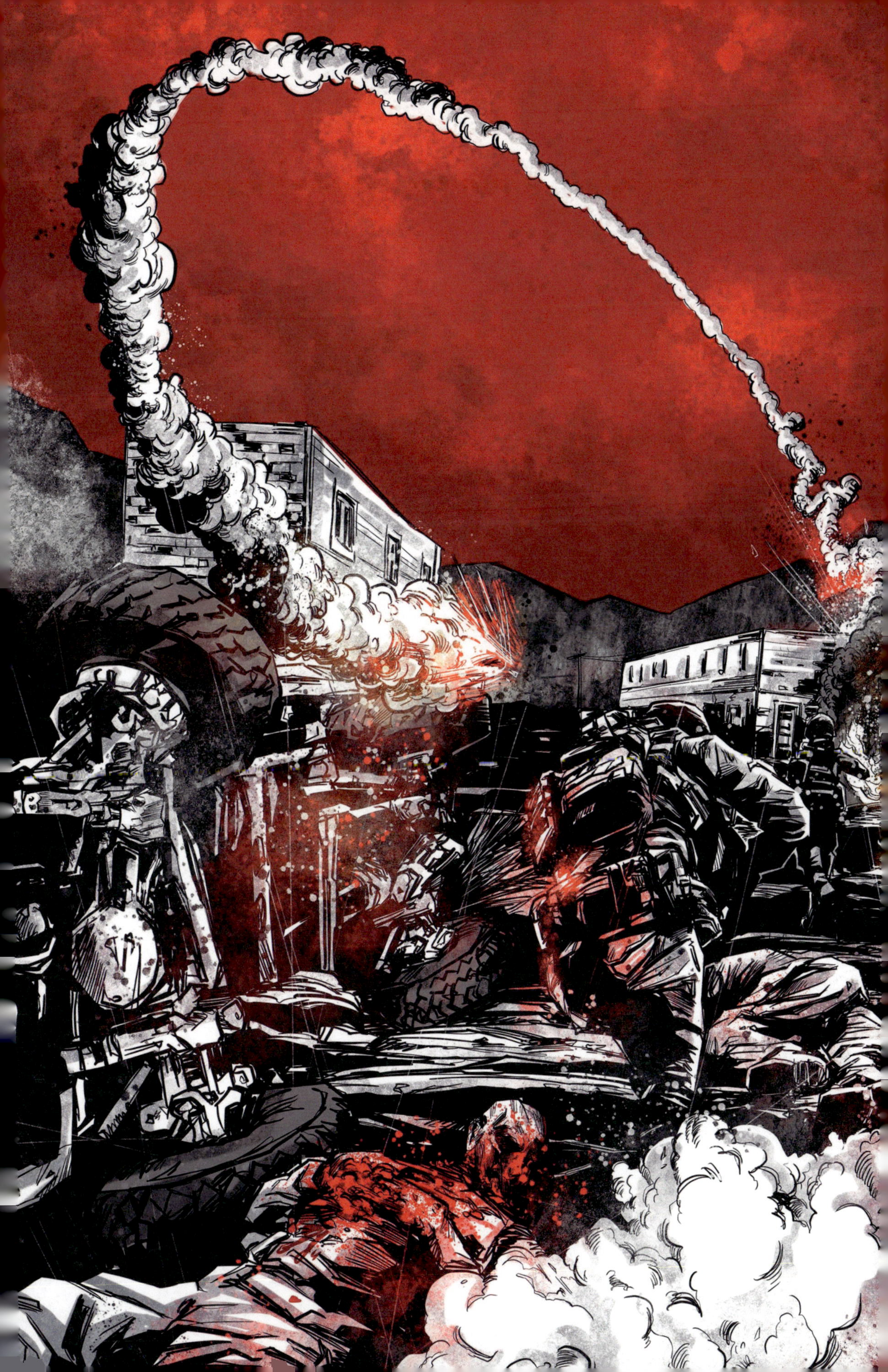

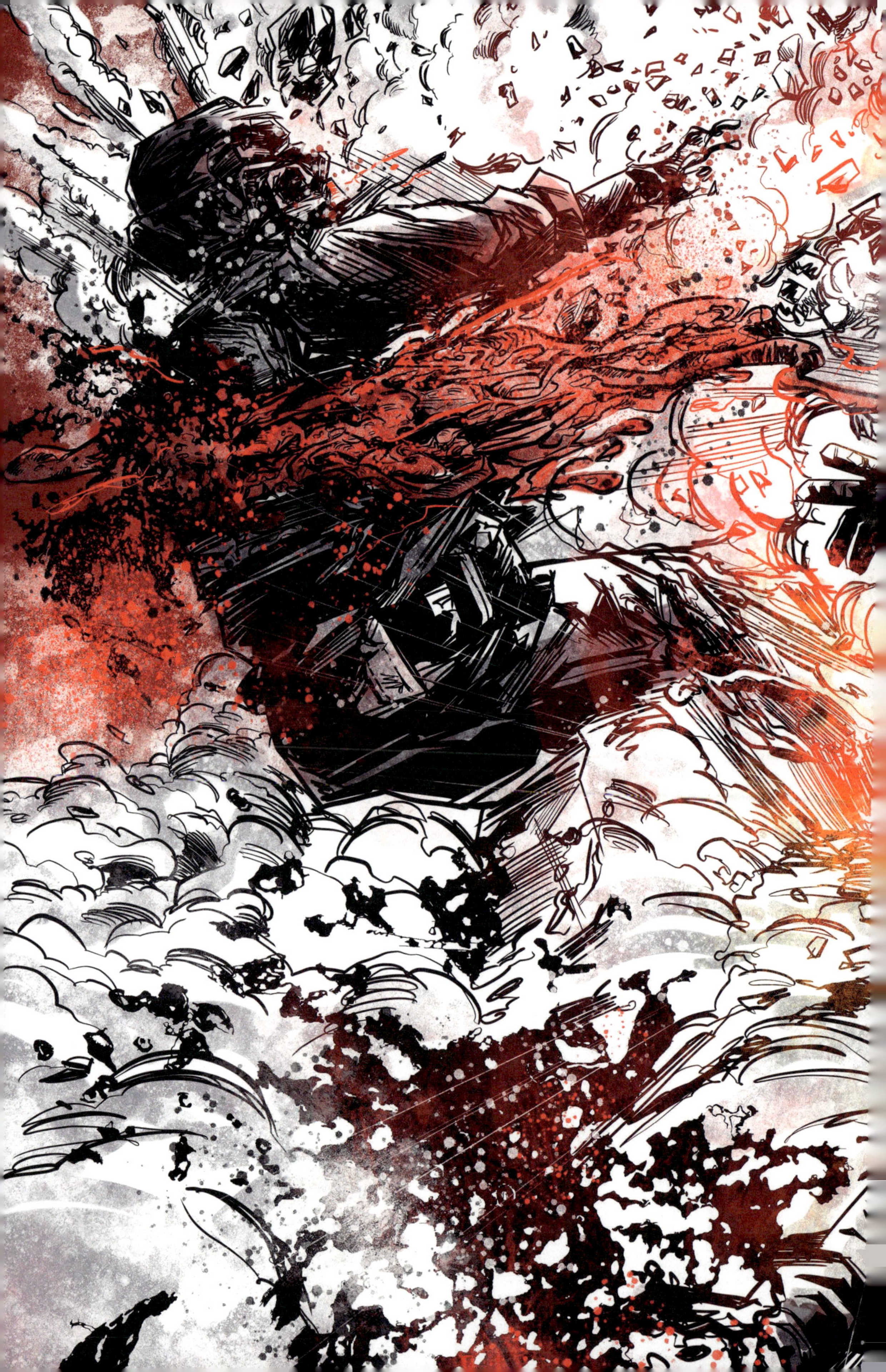

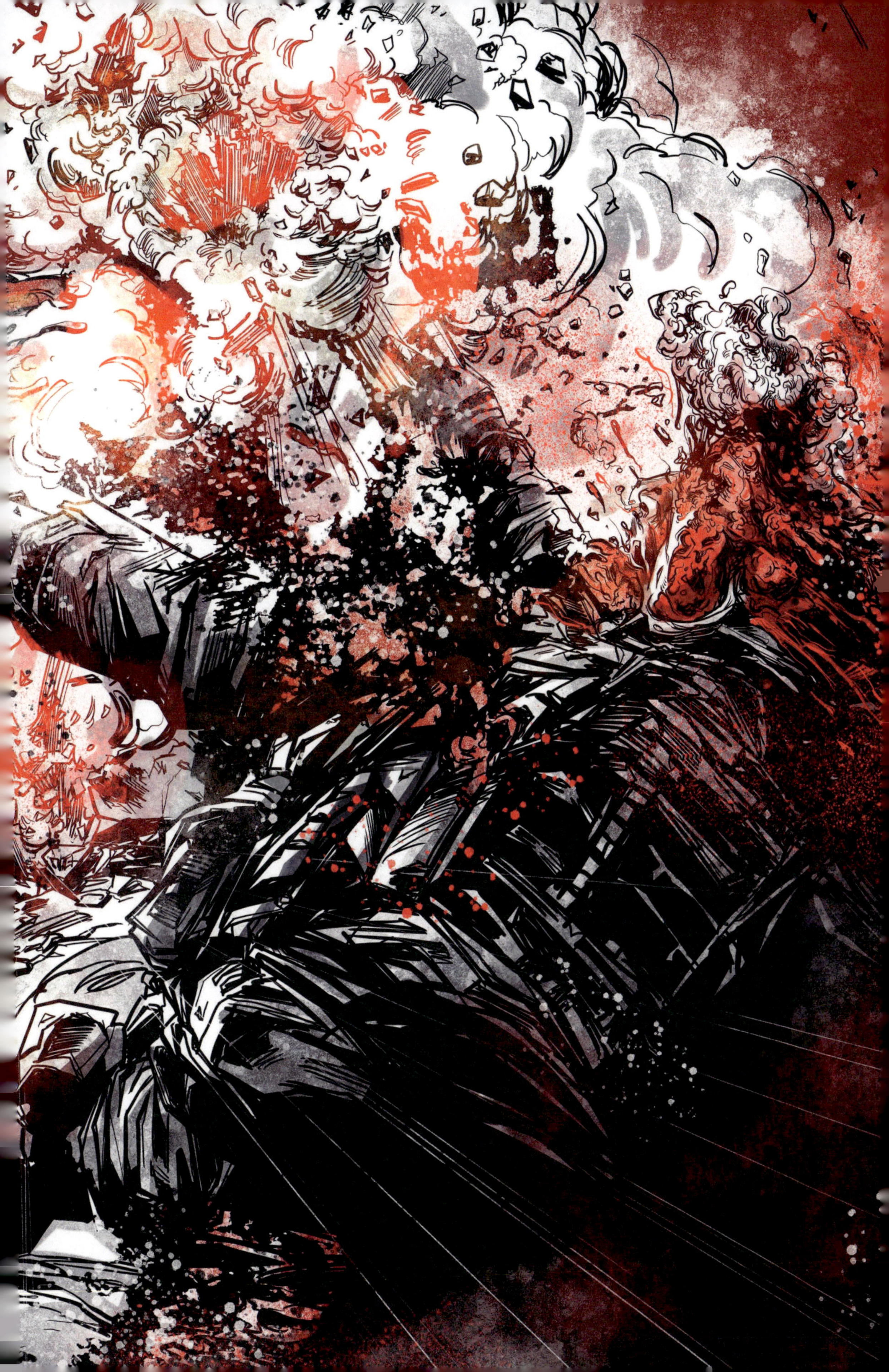

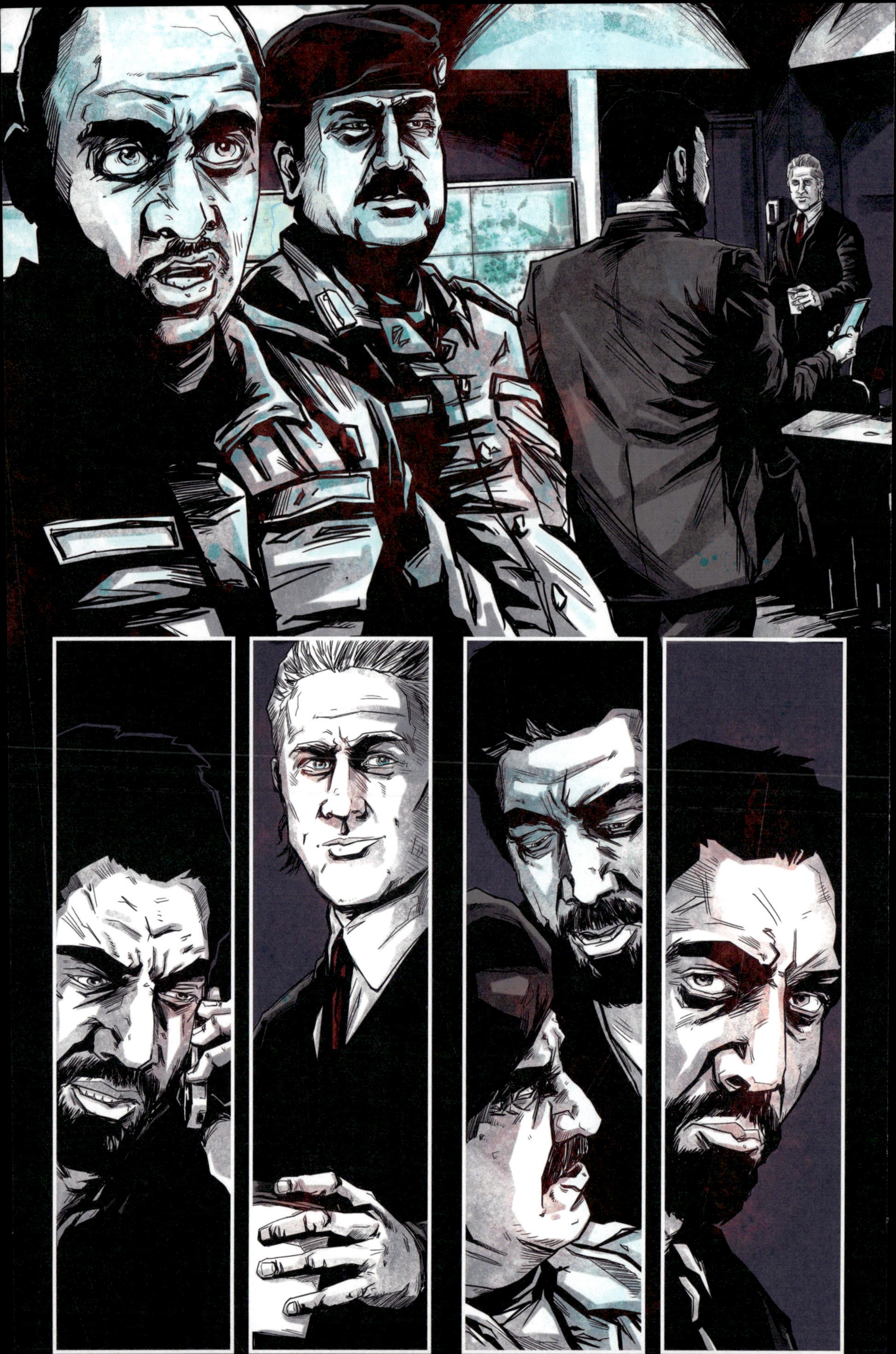

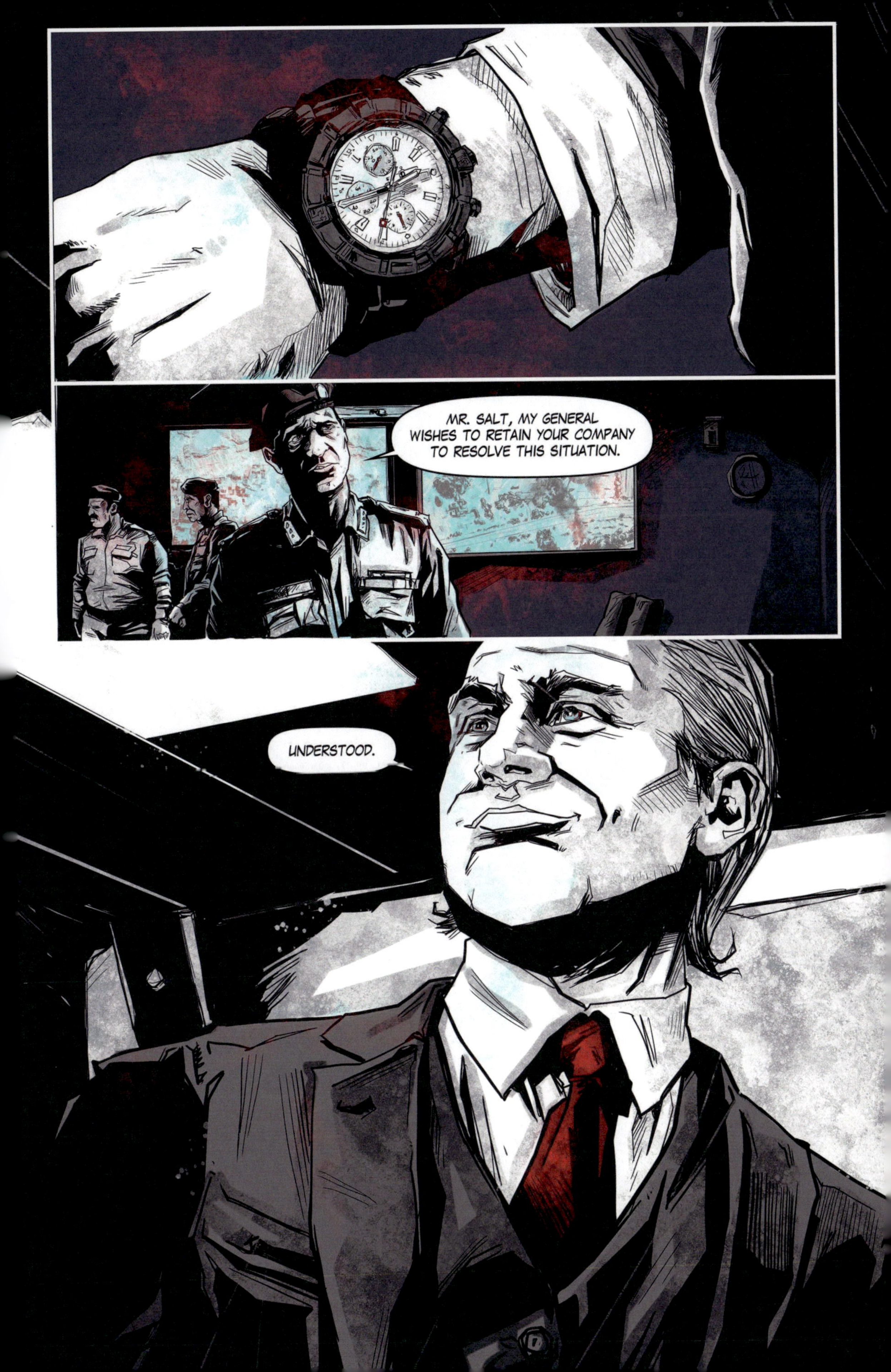
MR. SALT, MY GENERAL WISHES TO RETAIN YOUR COMPANY TO RESOLVE THIS SITUATION.
UNDERSTOOD.

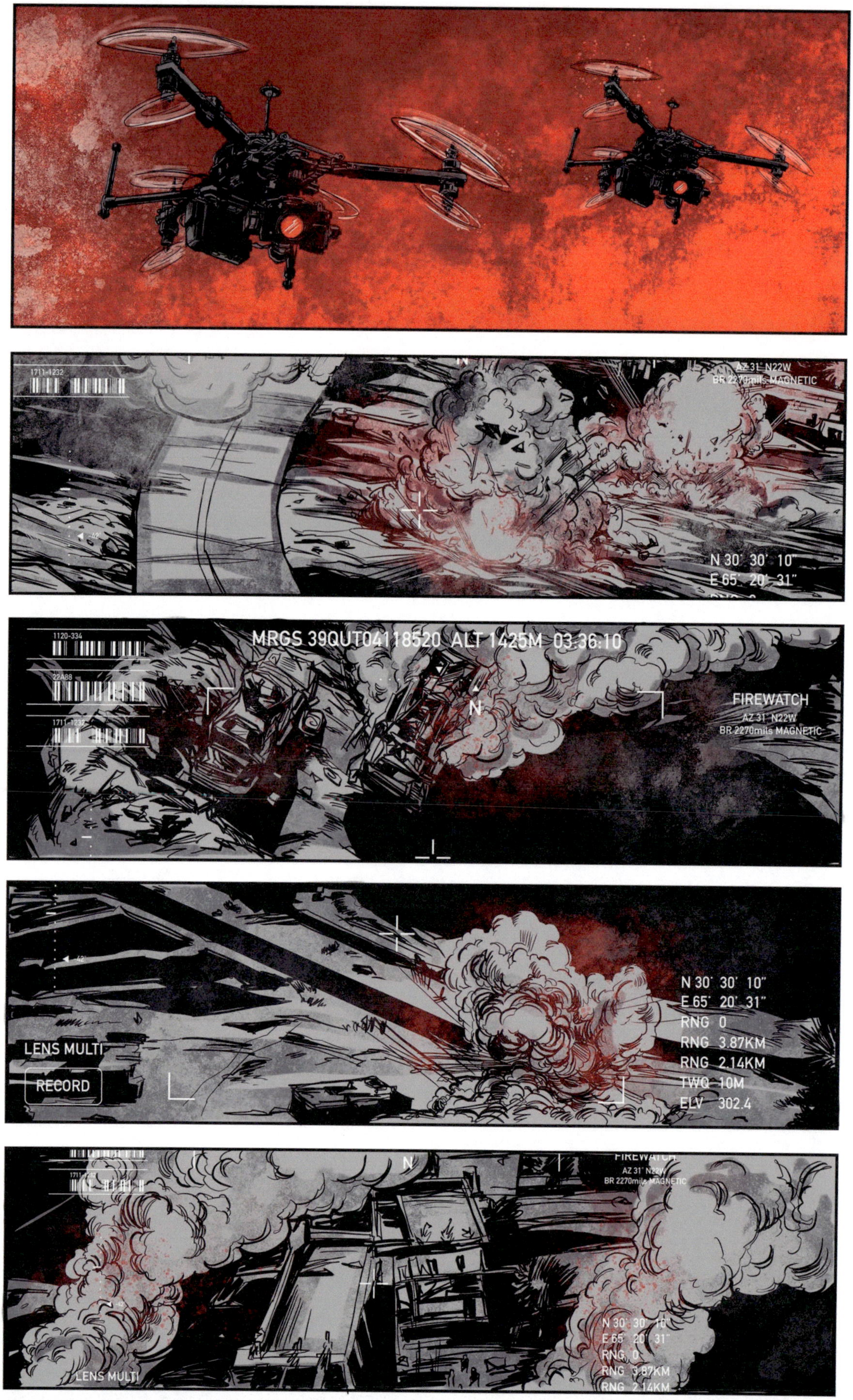
1711-1232
AZ 31' N22W
BR 2270mils MAGNETIC
N 30' 30' 10"
E 65' 20' 31"
1120-334
MRGS 39QUT04118520 ALT 1425M 03:36:10
22A88
1711-1232
N
FIREWATCH
AZ 31' N22W
BR 2270mils MAGNETIC
N 30' 30' 10"
E 65' 20' 31"
RNG 0
RNG 3.87KM
RNG 2.14KM
TWQ 10M
ELV 302.4
LENS MULTI
RECORD
N
AZ 31' N22W
BR 2270mils MAGNETIC
N 30' 30' 10"
E 65' 20' 31"
RNG 0
RNG 3.87KM
RNG 2.14KM
LENS MULTI

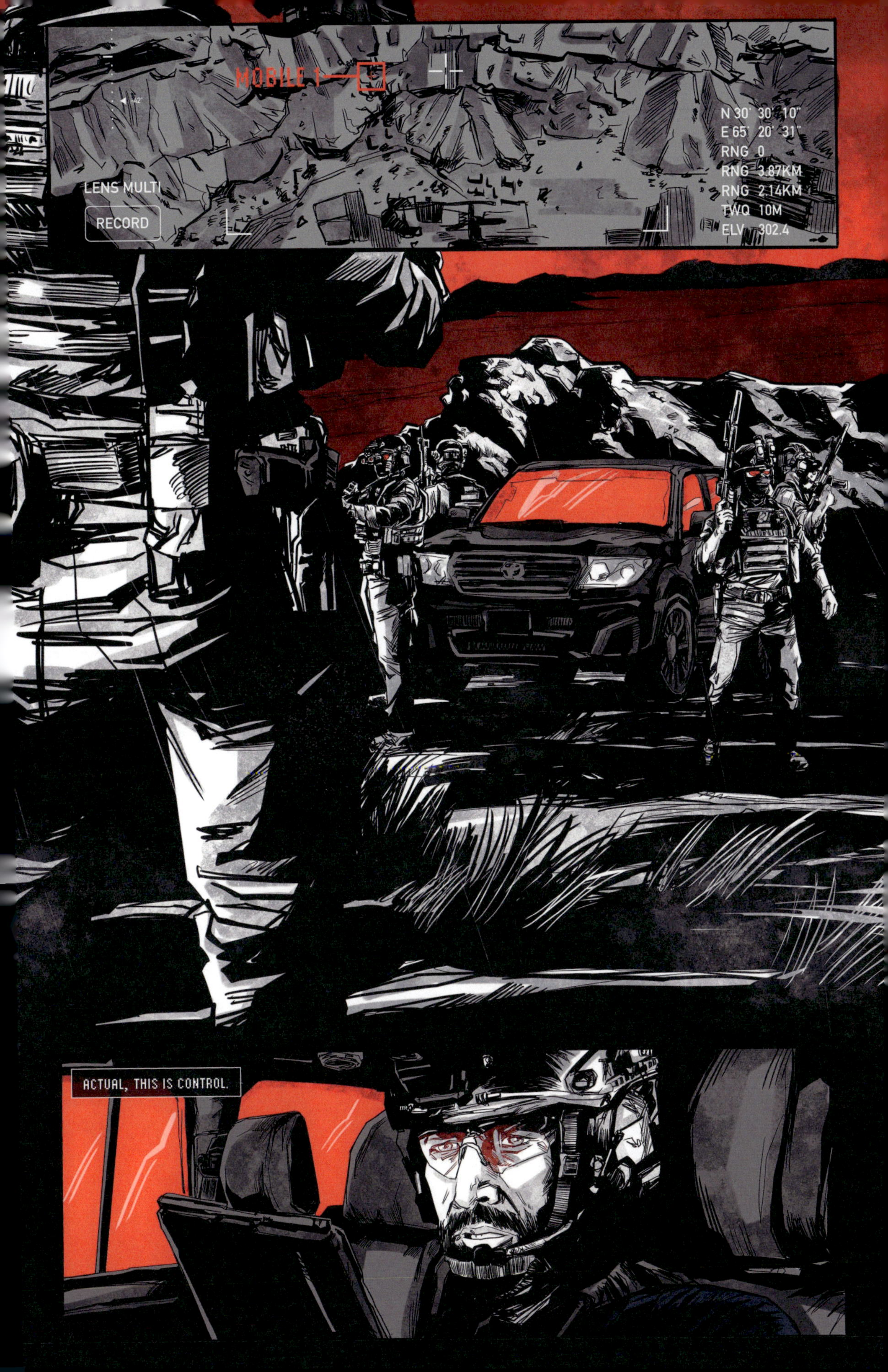
MOBILE 1
LENS MULTI
RECORD
N 30˙ 30' 10"
E 65˙ 20' 31"
RNG 0
RNG 3.87KM
RNG 2.14KM
TWQ 10M
ELV 302.4
ACTUAL, THIS IS CONTROL.

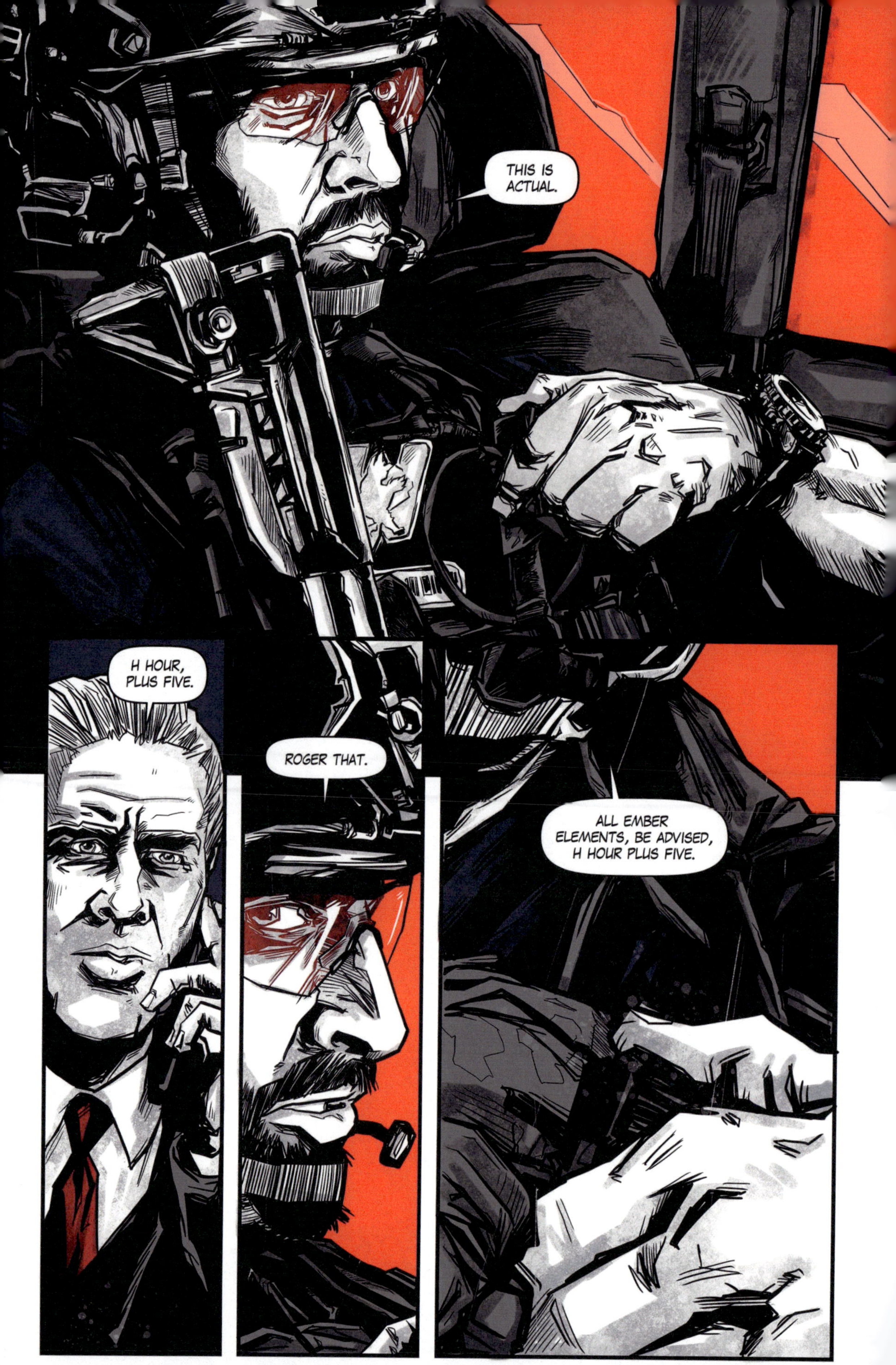
THIS IS ACTUAL.
H HOUR, PLUS FIVE.
ROGER THAT.
ALL EMBER ELEMENTS, BE ADVISED, H HOUR PLUS FIVE.

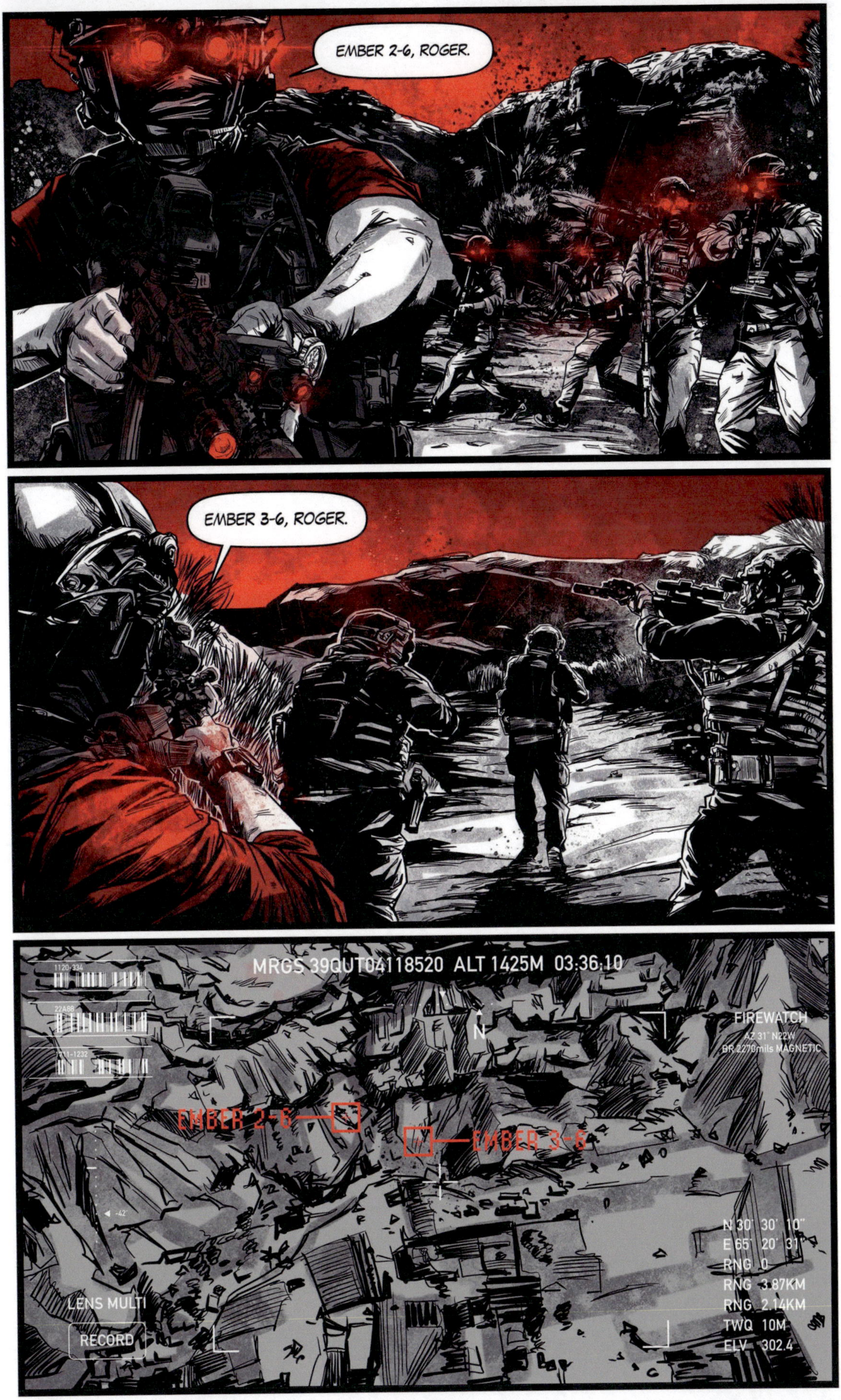
EMBER 2-6, ROGER.
EMBER 3-6, ROGER.
MRGS 39QUT04118520 ALT 1425M 03:36:10
1120-334
22A88
1711-1232
N
FIREWATCH
AZ 31° N22W
BR 2270mils MAGNETIC
EMBER 2-6
EMBER 3-6
N 30° 30' 10"
E 65° 20' 31"
RNG 0
RNG 3.87KM
RNG 2.14KM
TWQ 10M
ELV 302.4
LENS MULTI
RECORD

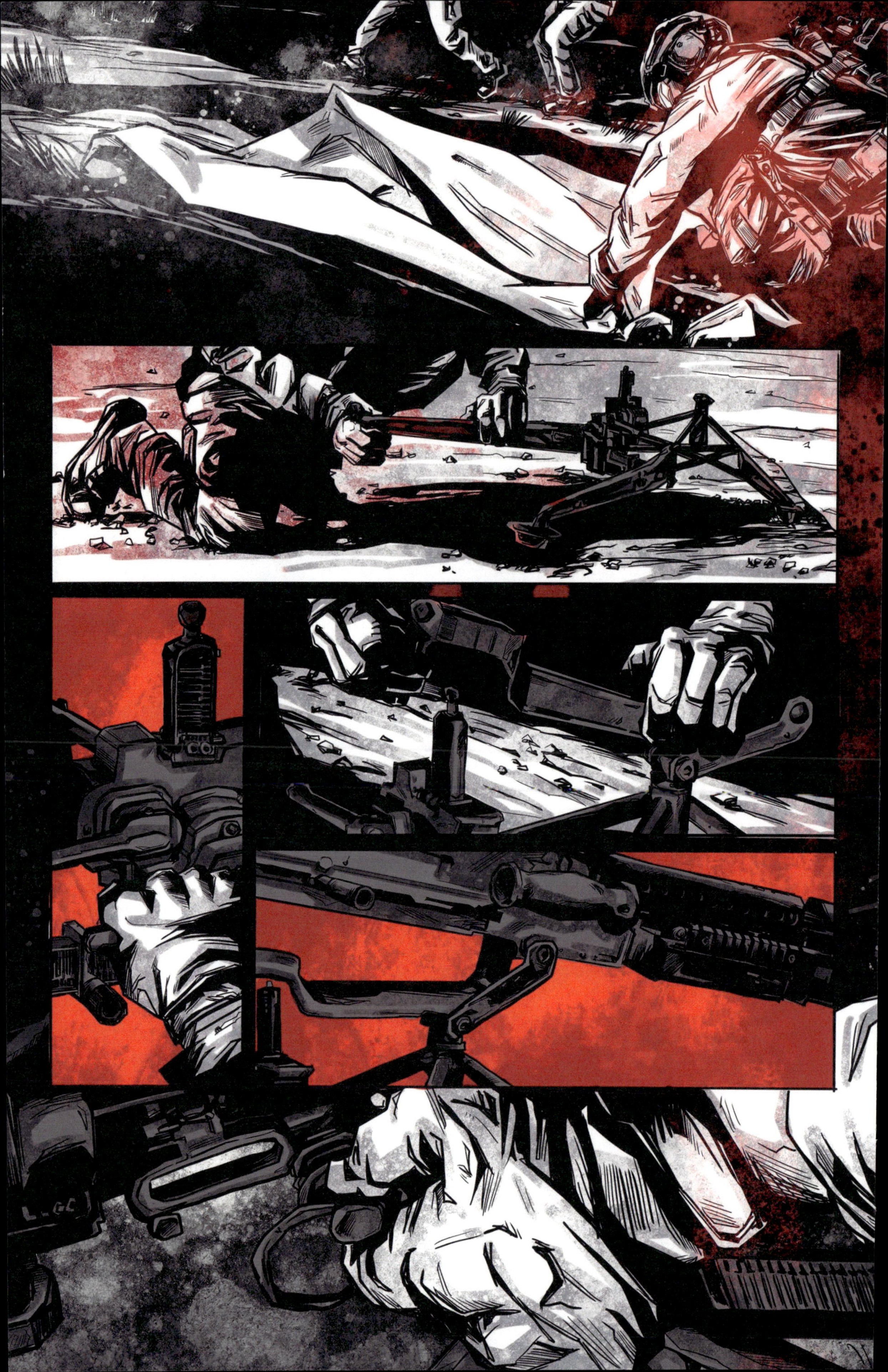

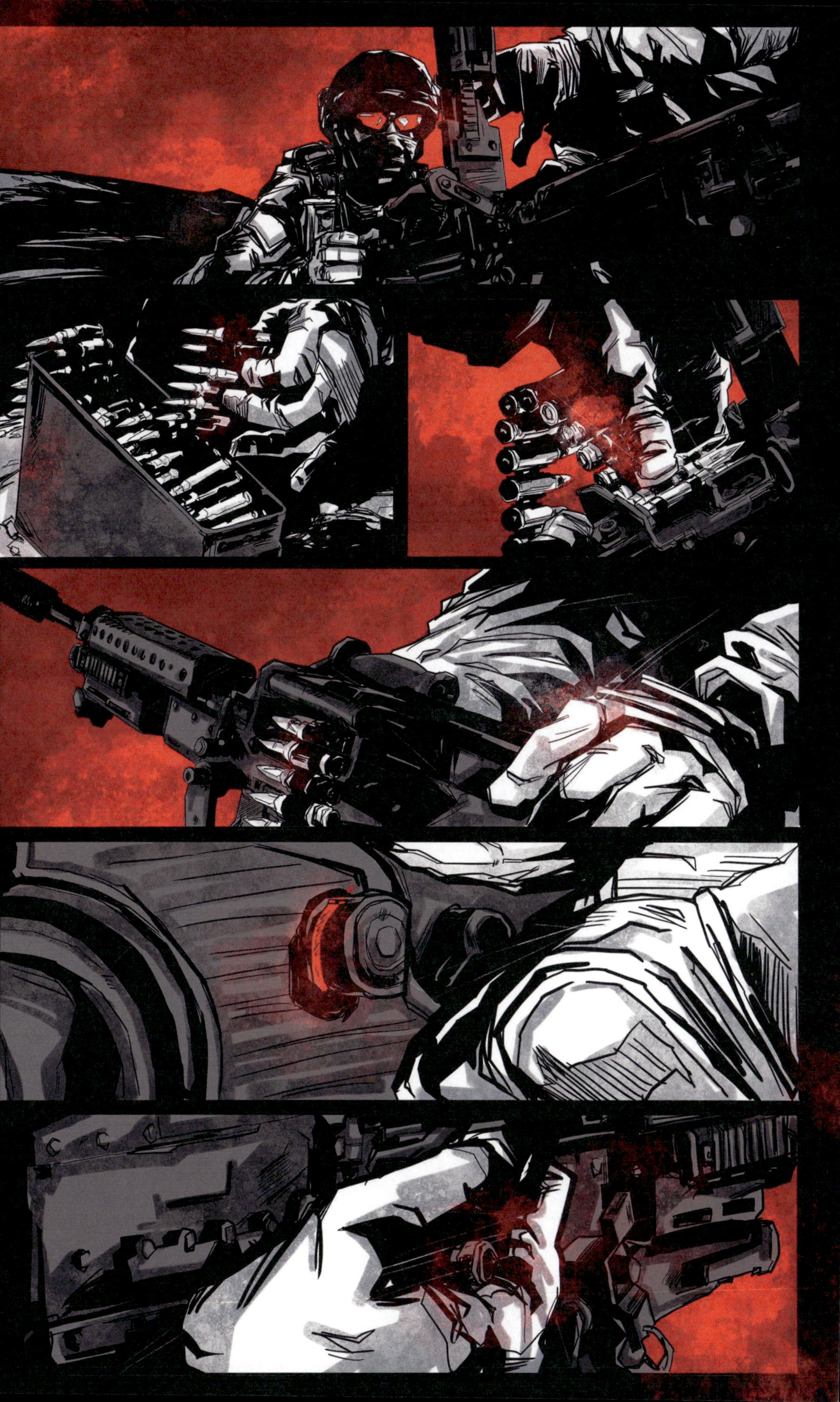

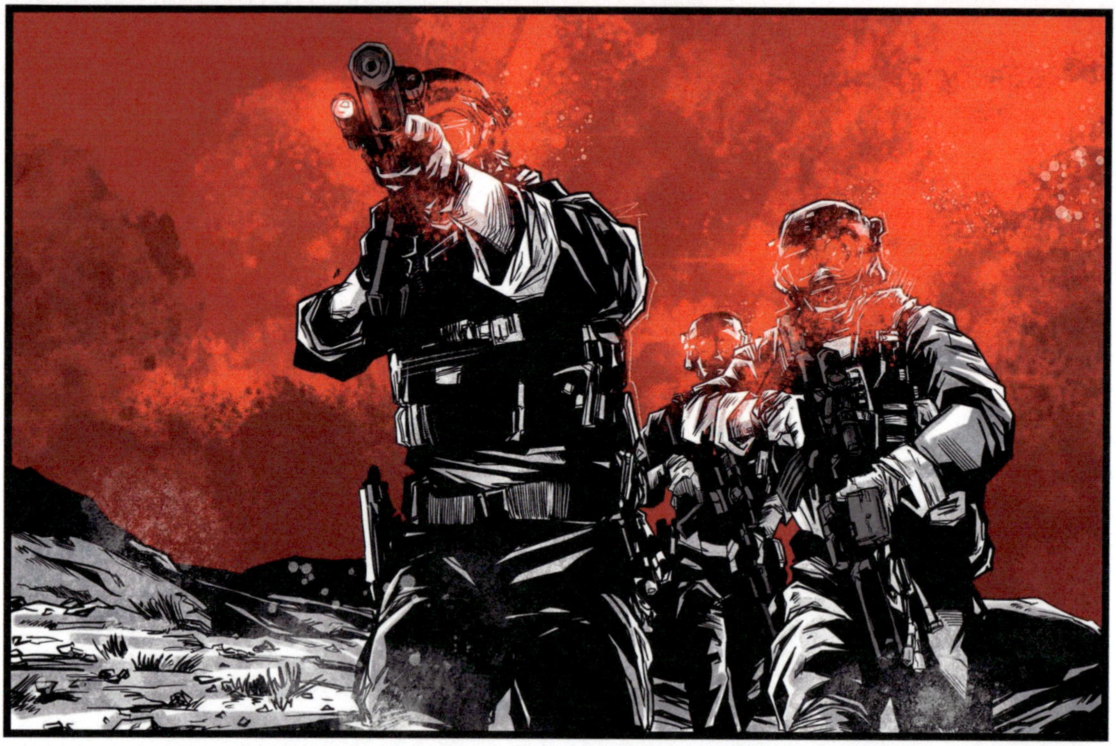

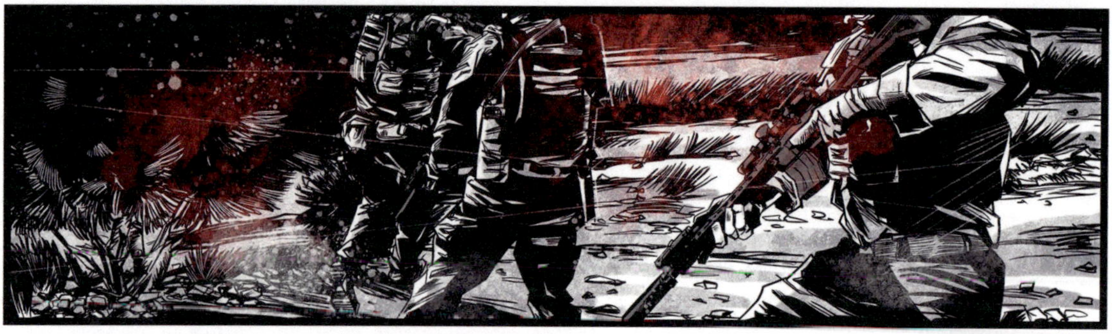

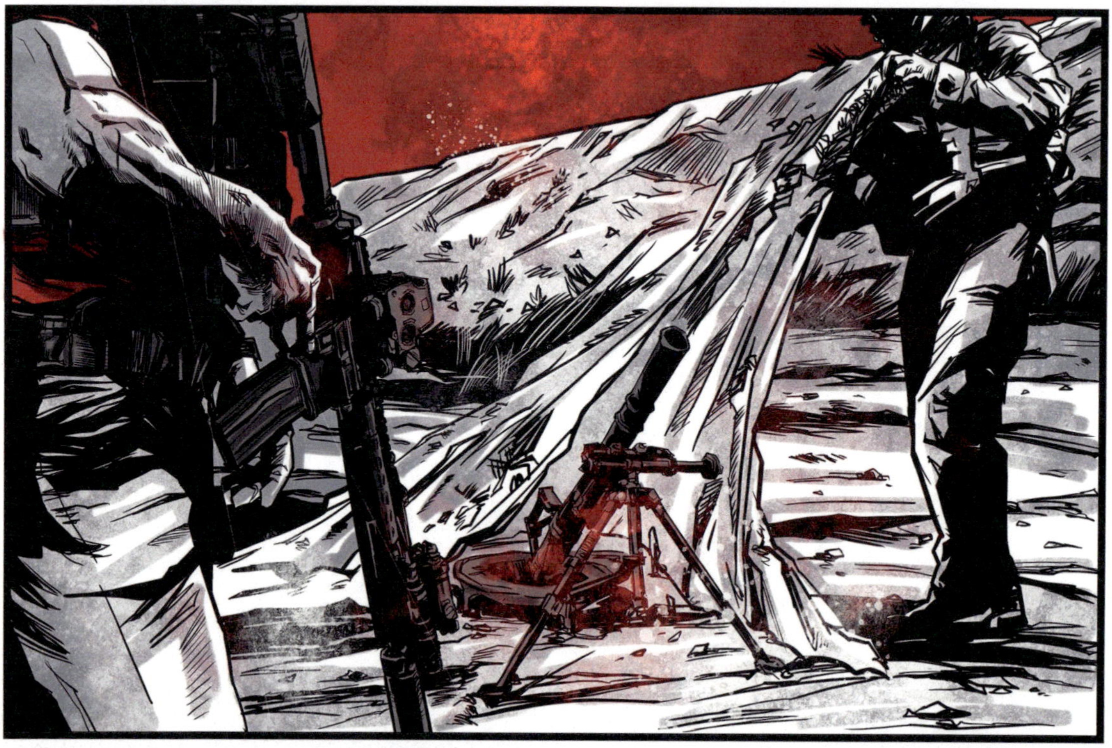

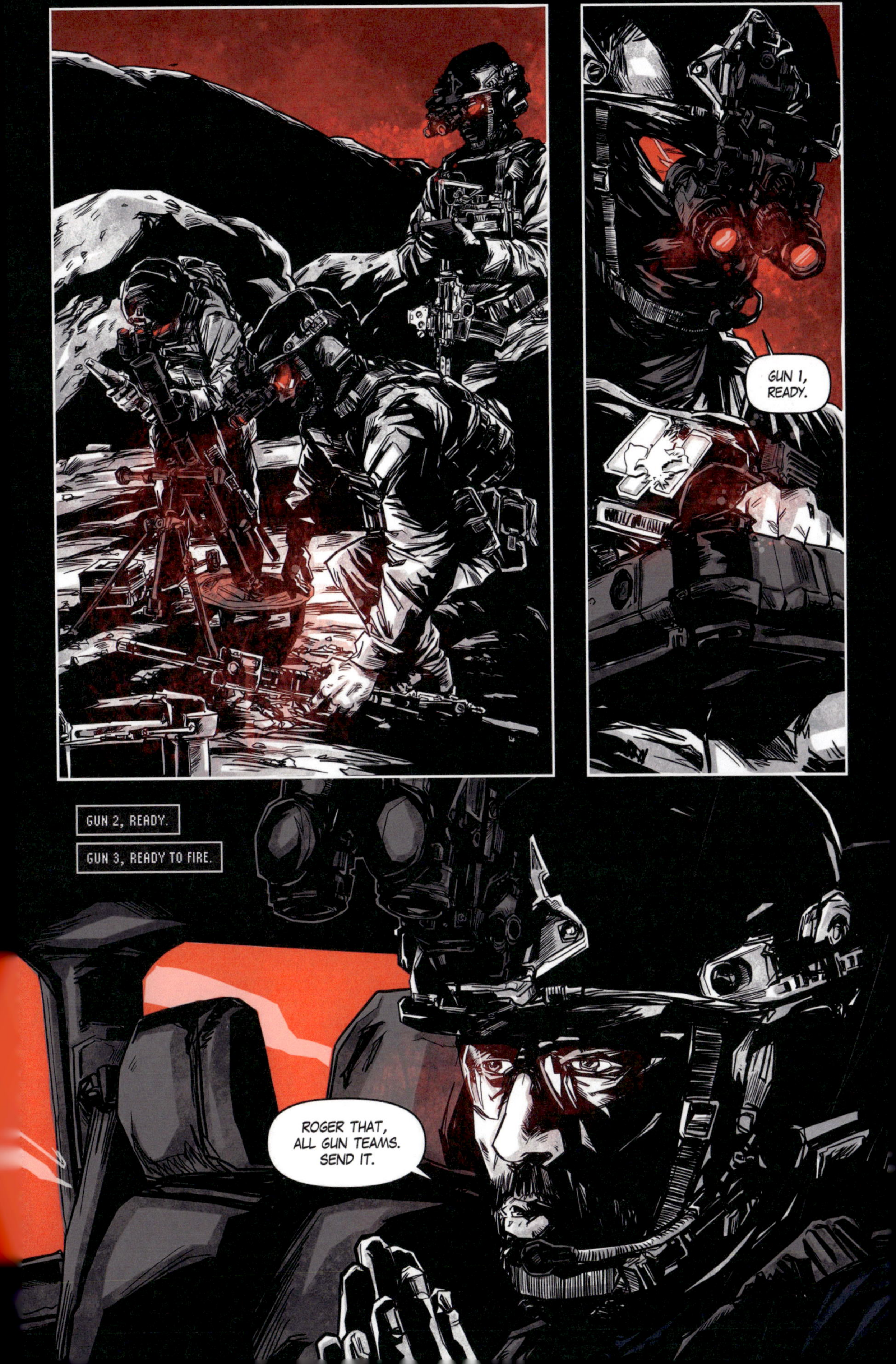
GUN 1, READY.
GUN 2, READY.
GUN 3, READY TO FIRE.
ROGER THAT, ALL GUN TEAMS. SEND IT.

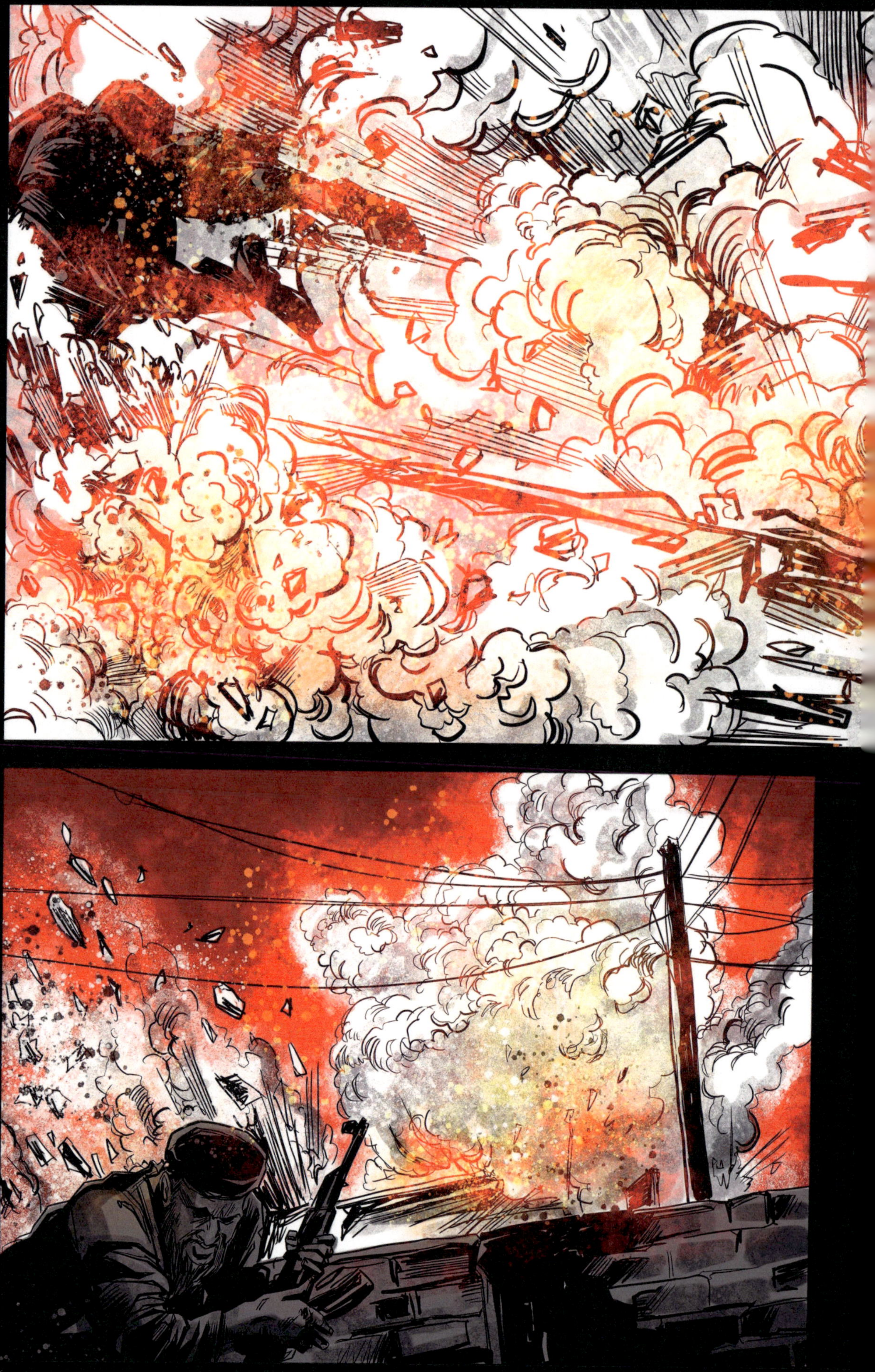

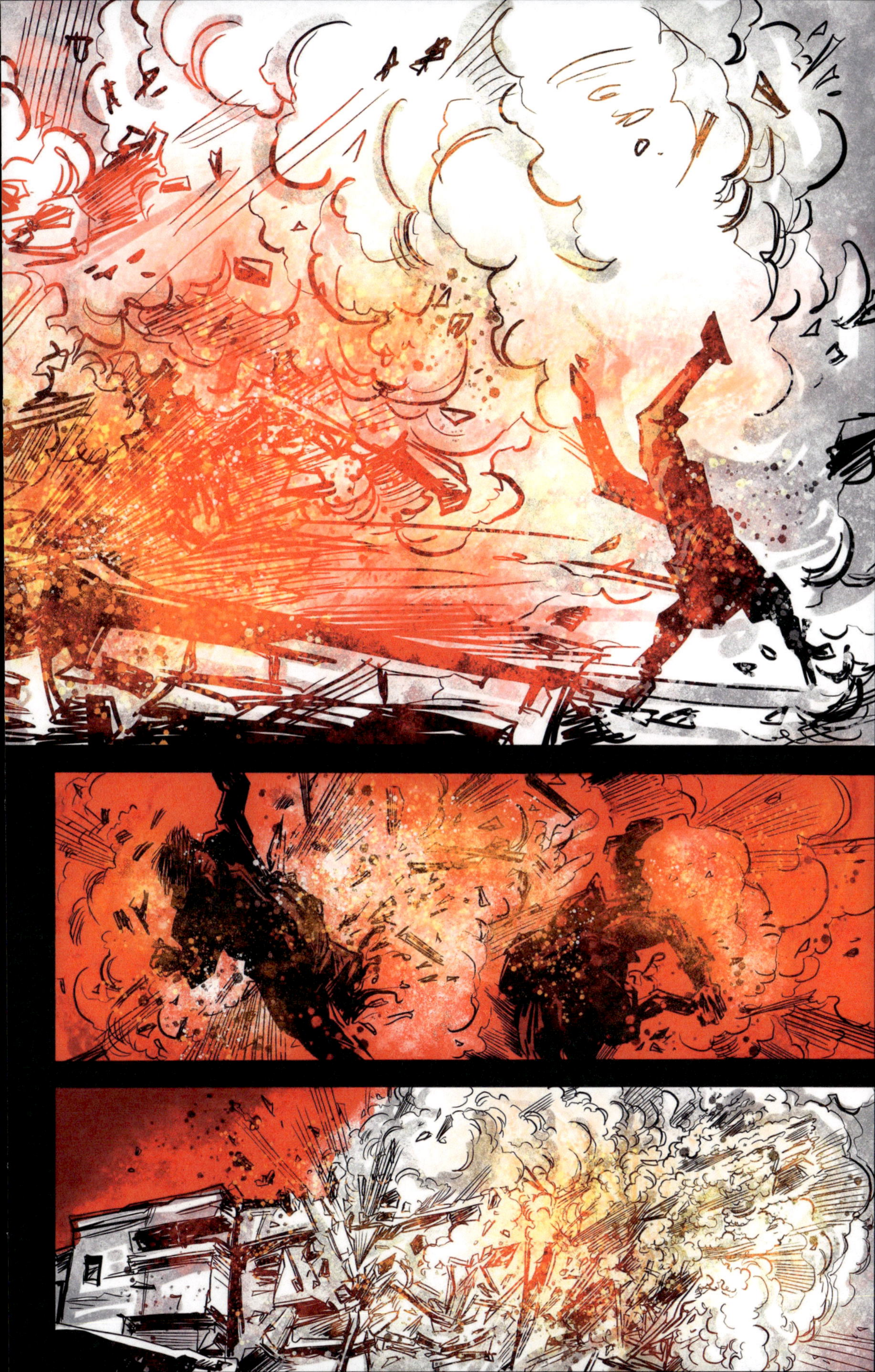

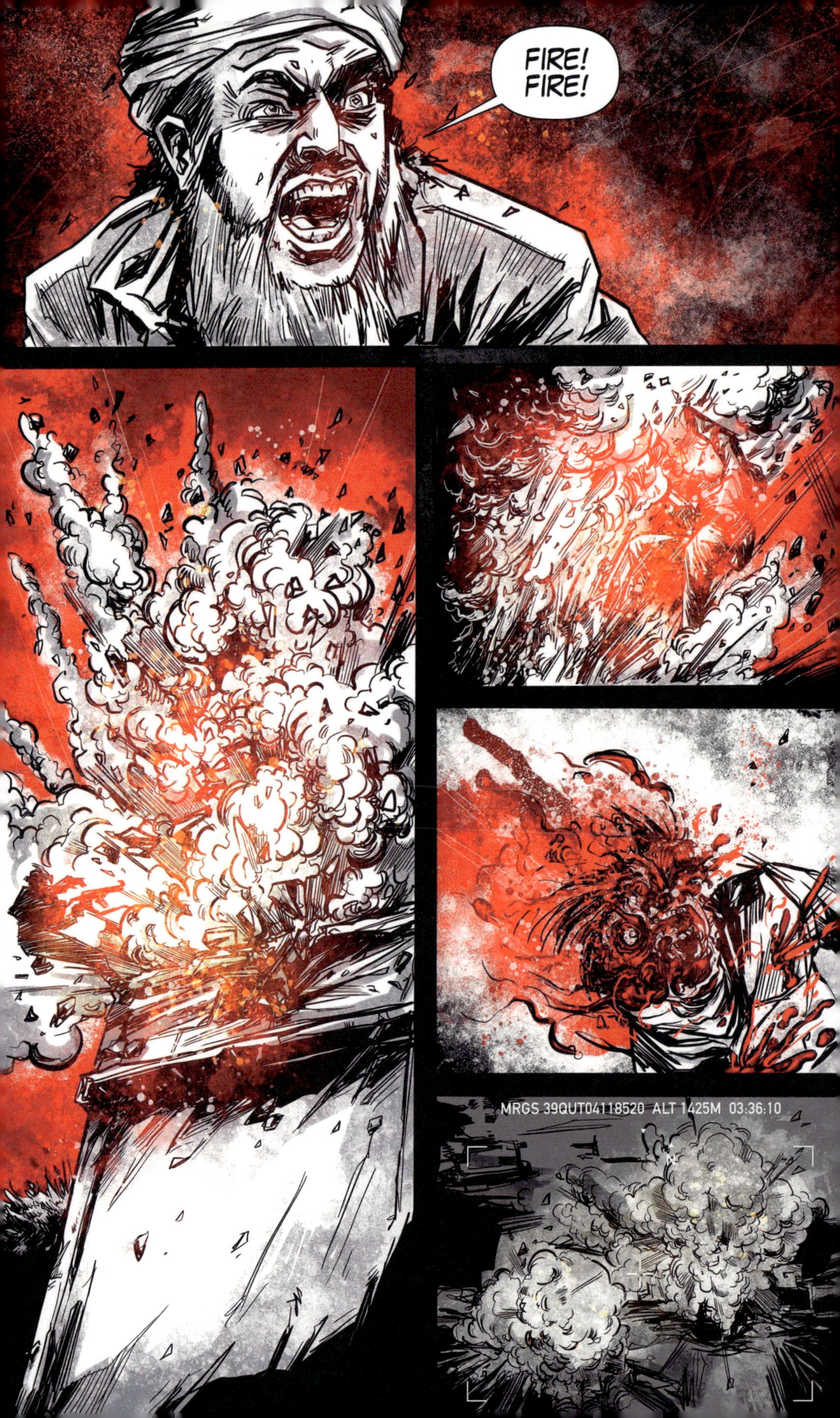
FIRE!
FIRE!
MRGS 39QUT04118520 ALT 1425M 03:36:10

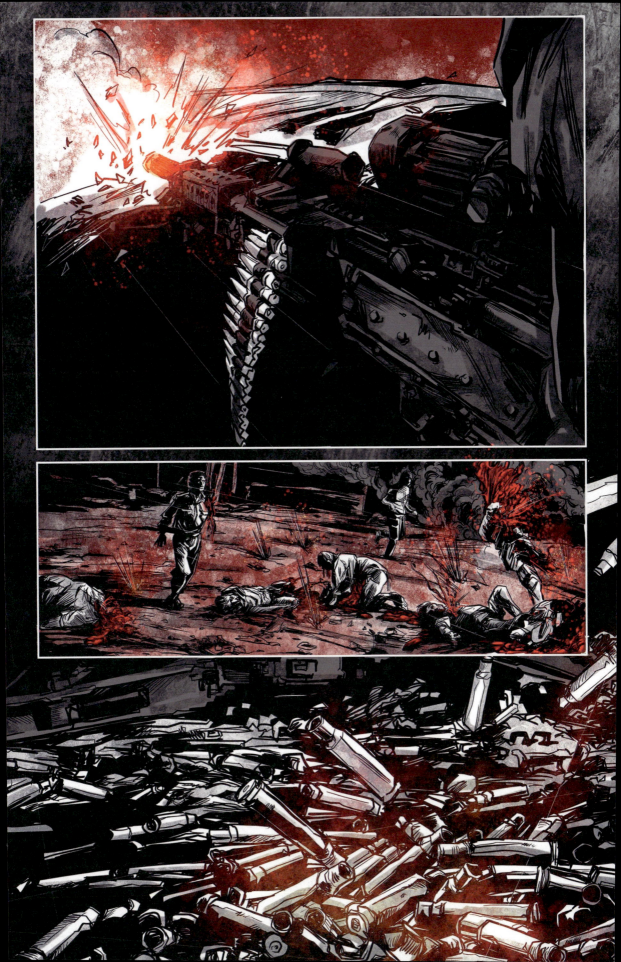

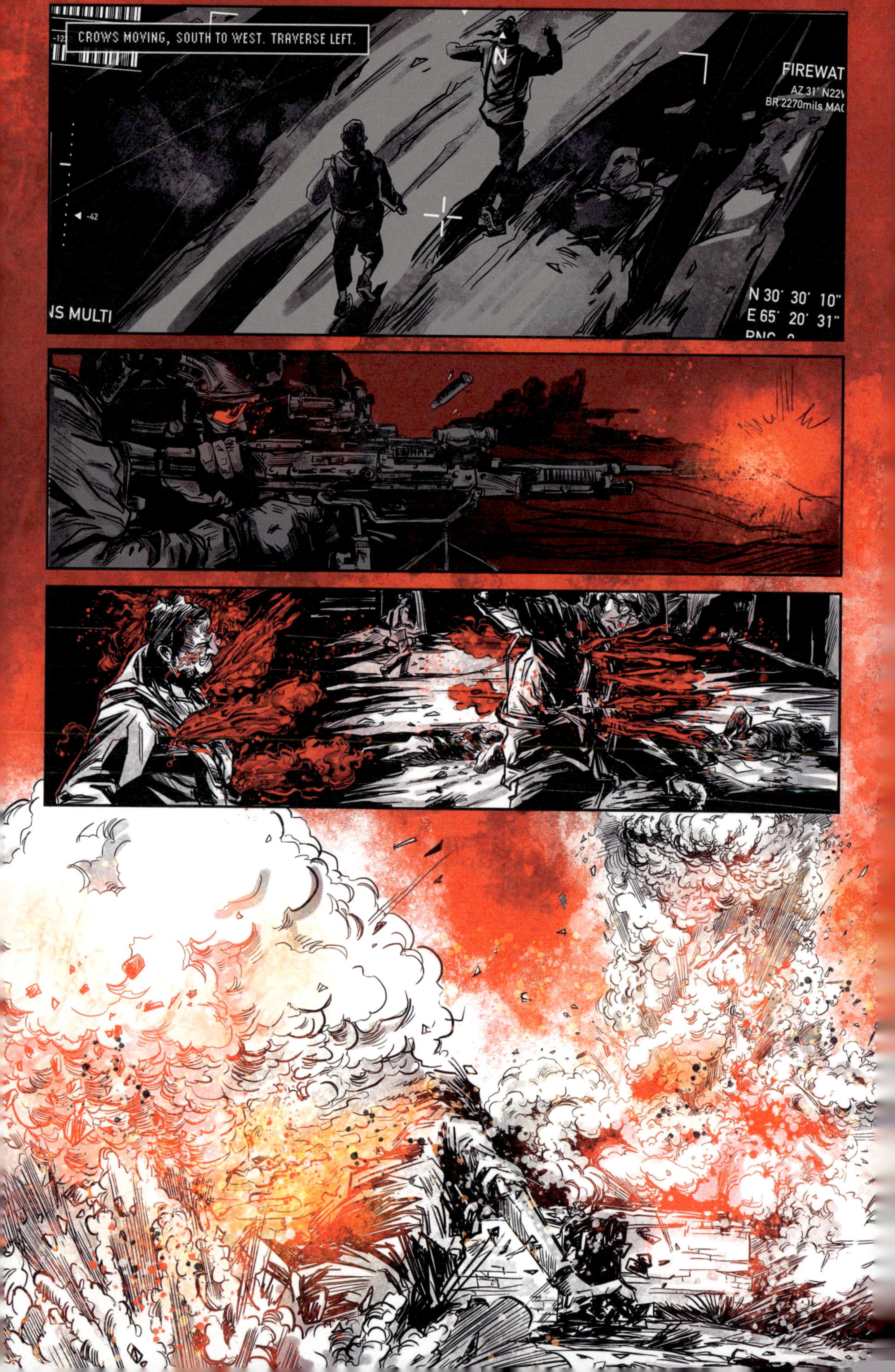
CROWS MOVING, SOUTH TO WEST. TRAVERSE LEFT.
FIREWAT
AZ 31° N22V
BR 2270mils MAC
-42
NS MULTI
N 30° 30' 10"
E 65° 20' 31"

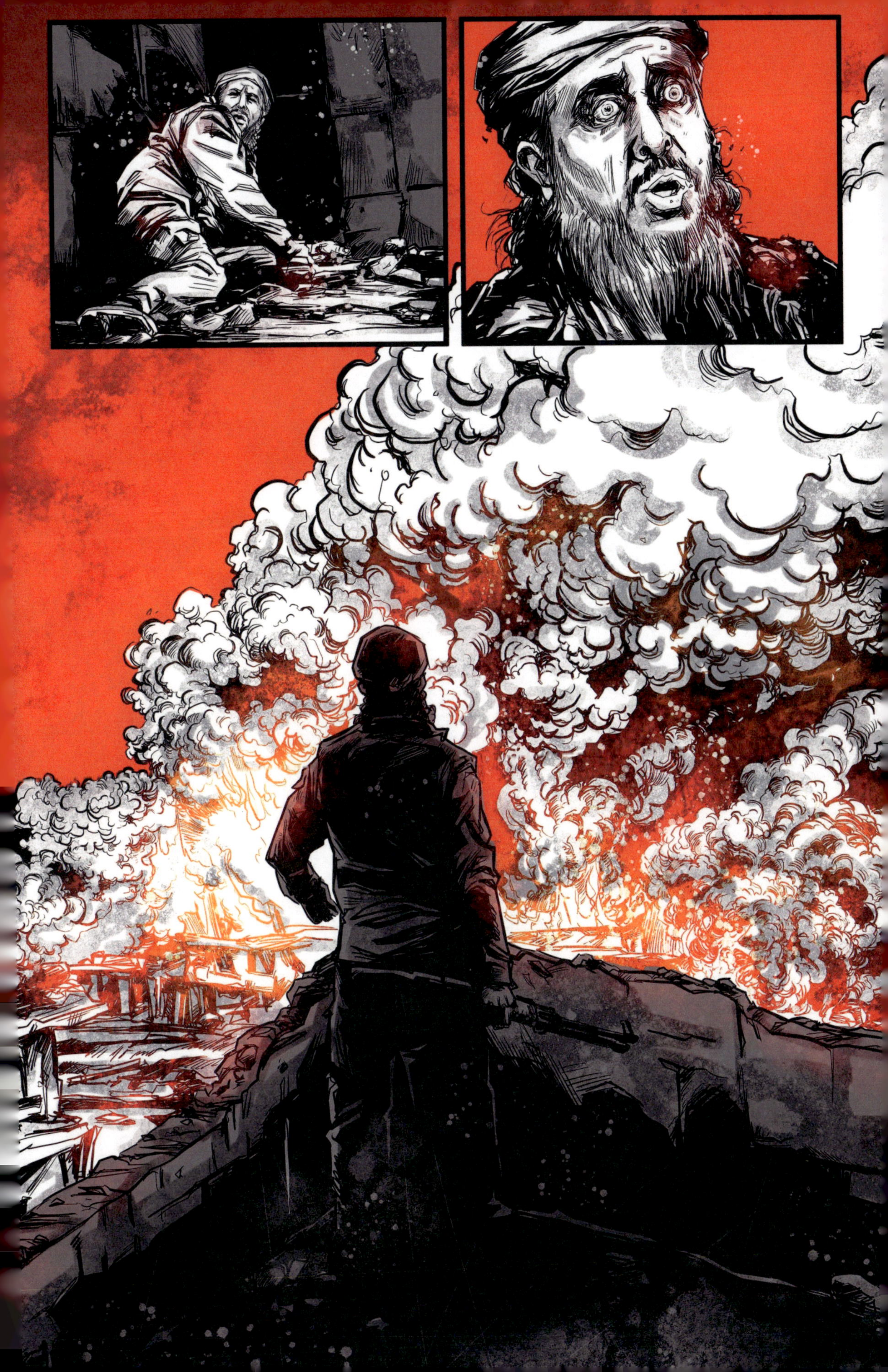

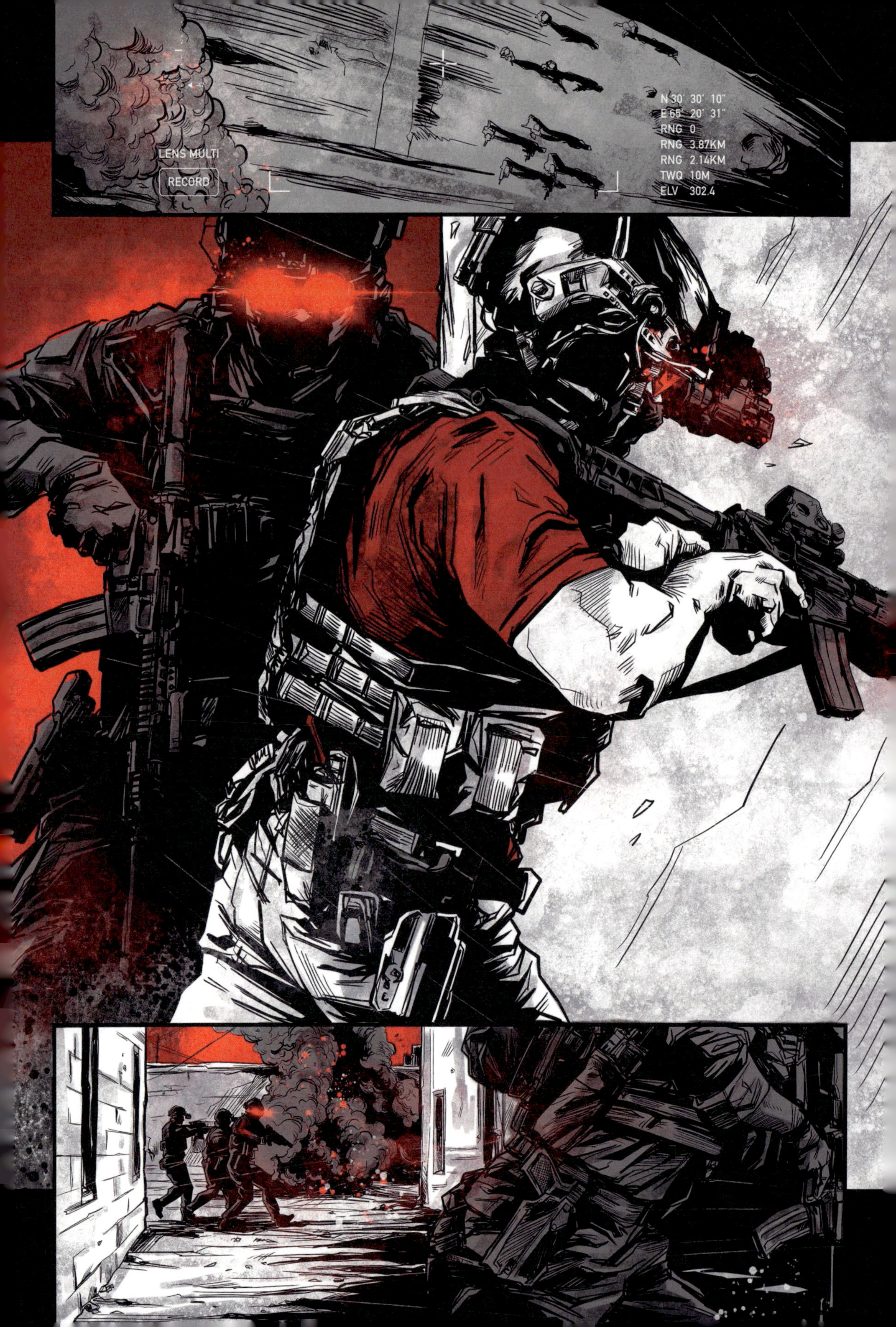
LENS MULTI
RECORD
N 30' 30' 10"
E 65' 20' 31"
RNG 0
RNG 3.87KM
RNG 2.14KM
TWQ 10M
ELV 302.4

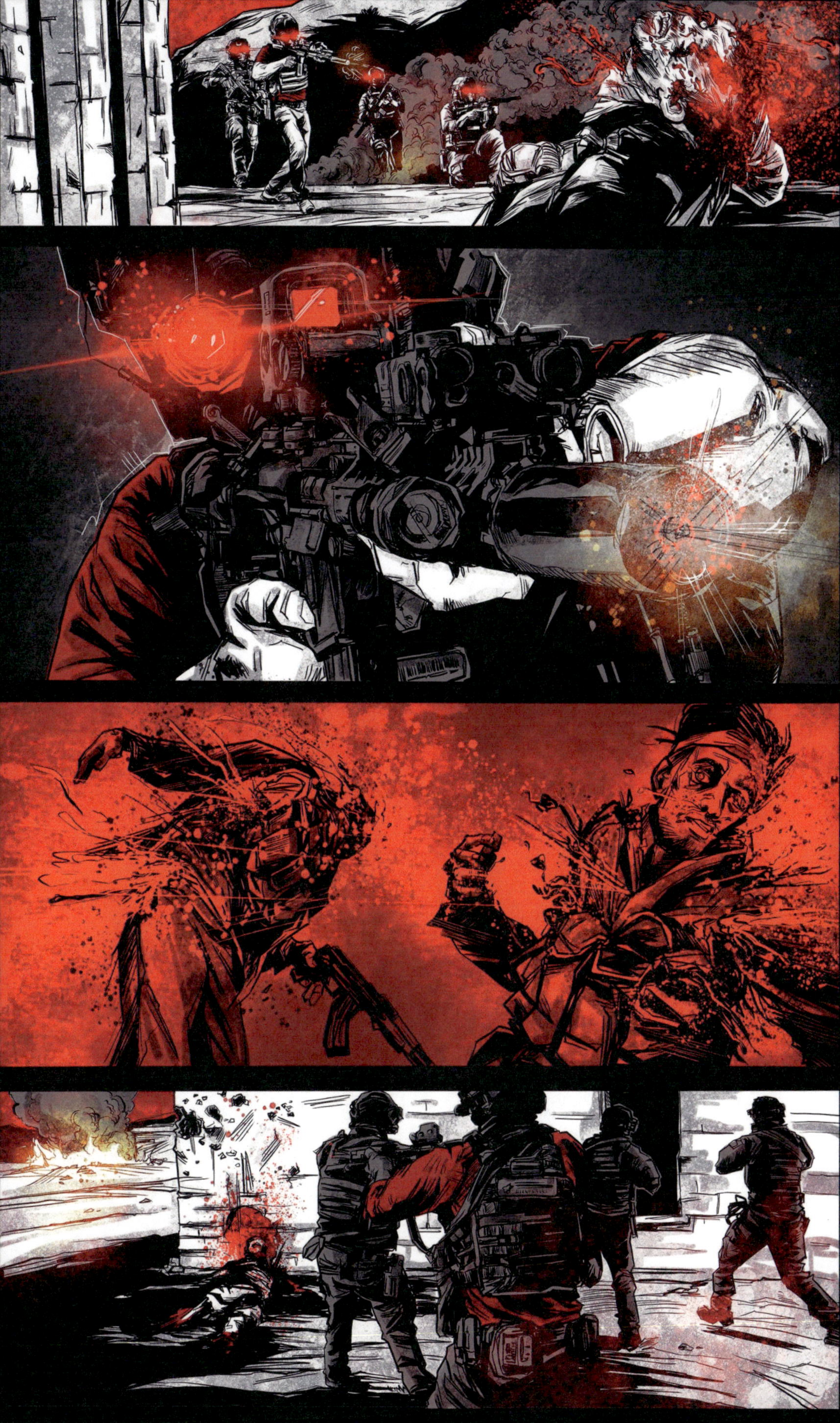

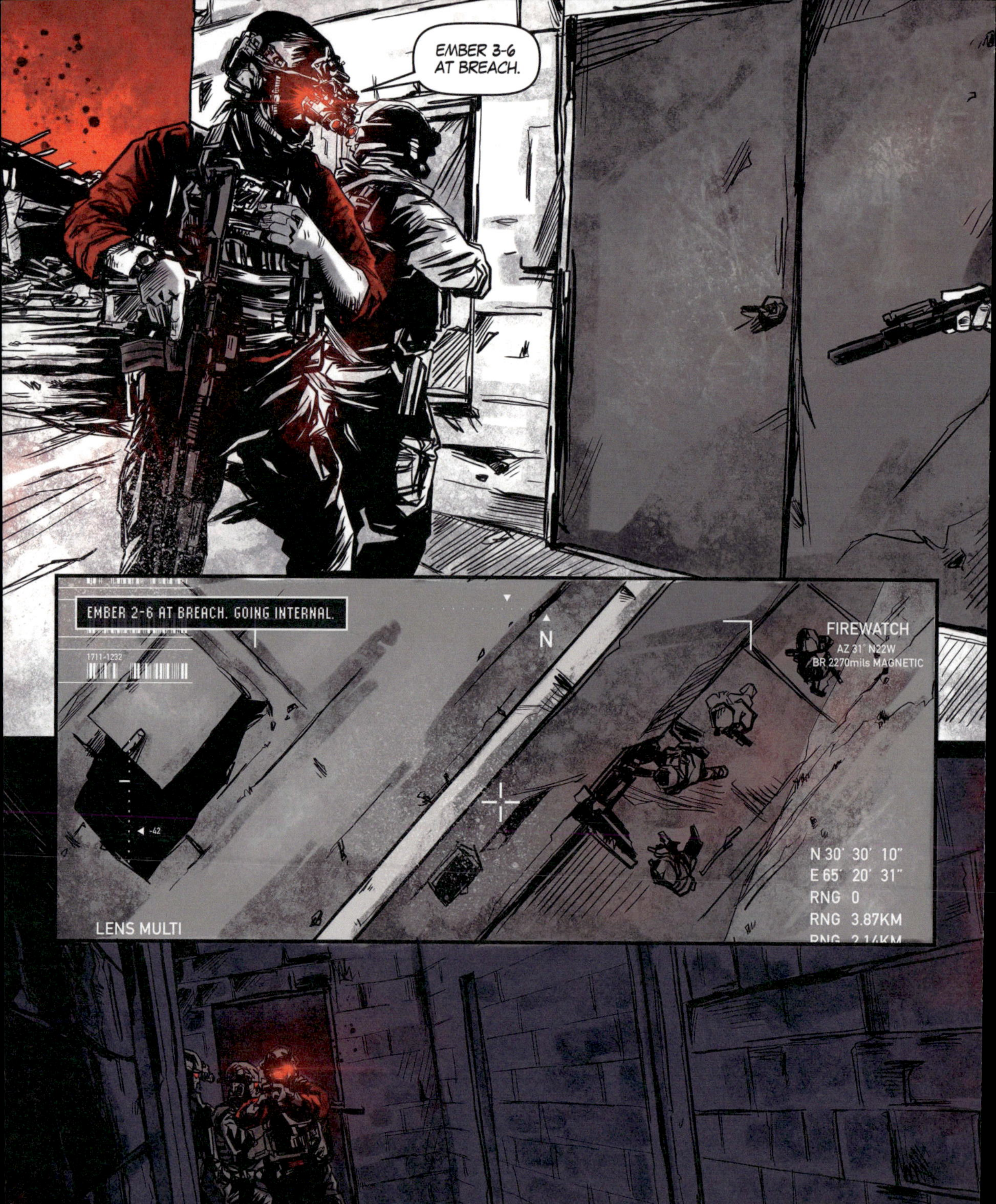
EMBER 3-6 AT BREACH.
EMBER 2-6 AT BREACH. GOING INTERNAL.
N
FIREWATCH
AZ 31° N22W
BR 2270mils MAGNETIC
N 30' 30' 10"
E 65' 20' 31"
RNG 0
RNG 3.87KM
LENS MULTI

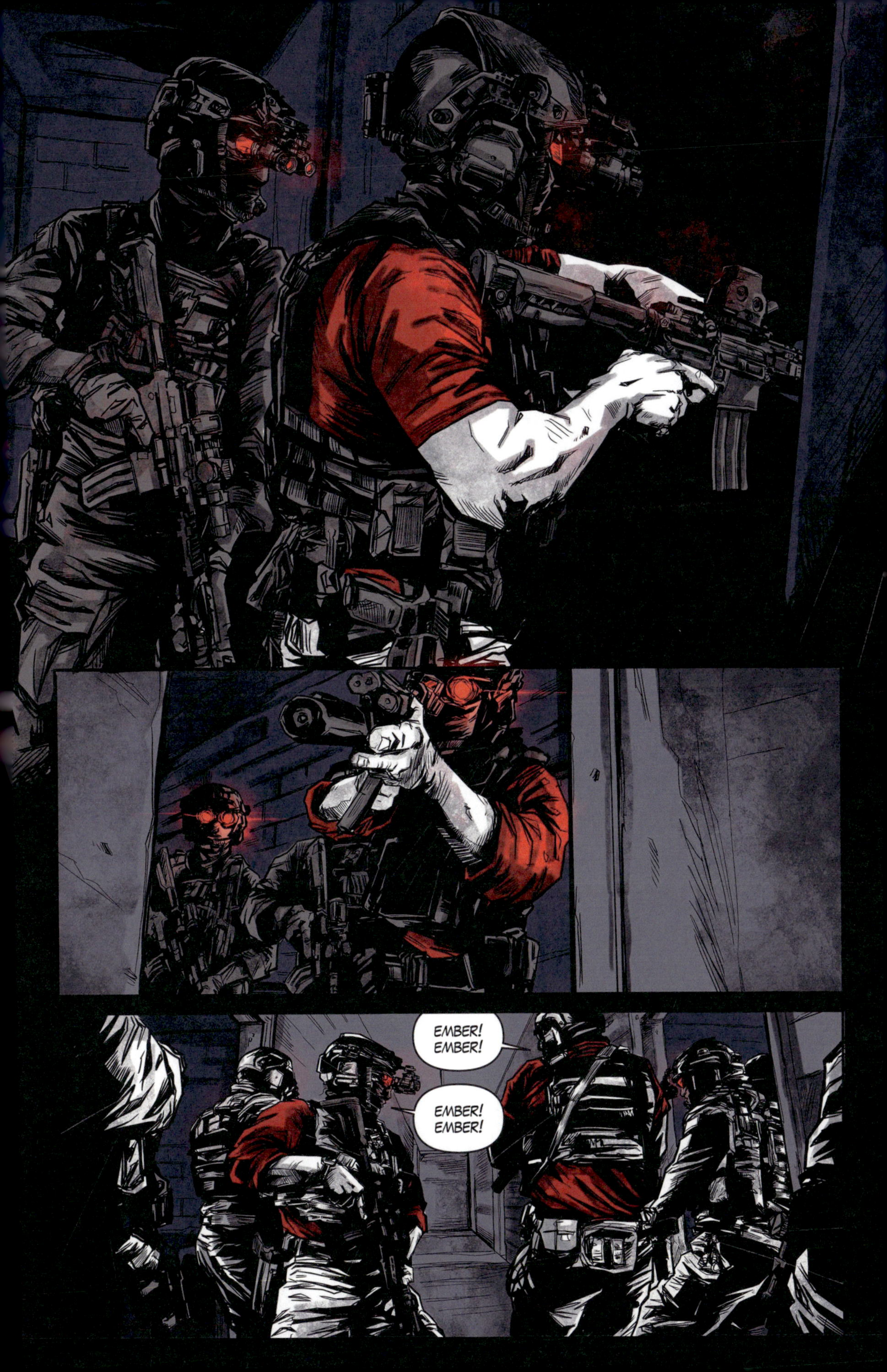
EMBER!
EMBER!
EMBER!
EMBER!

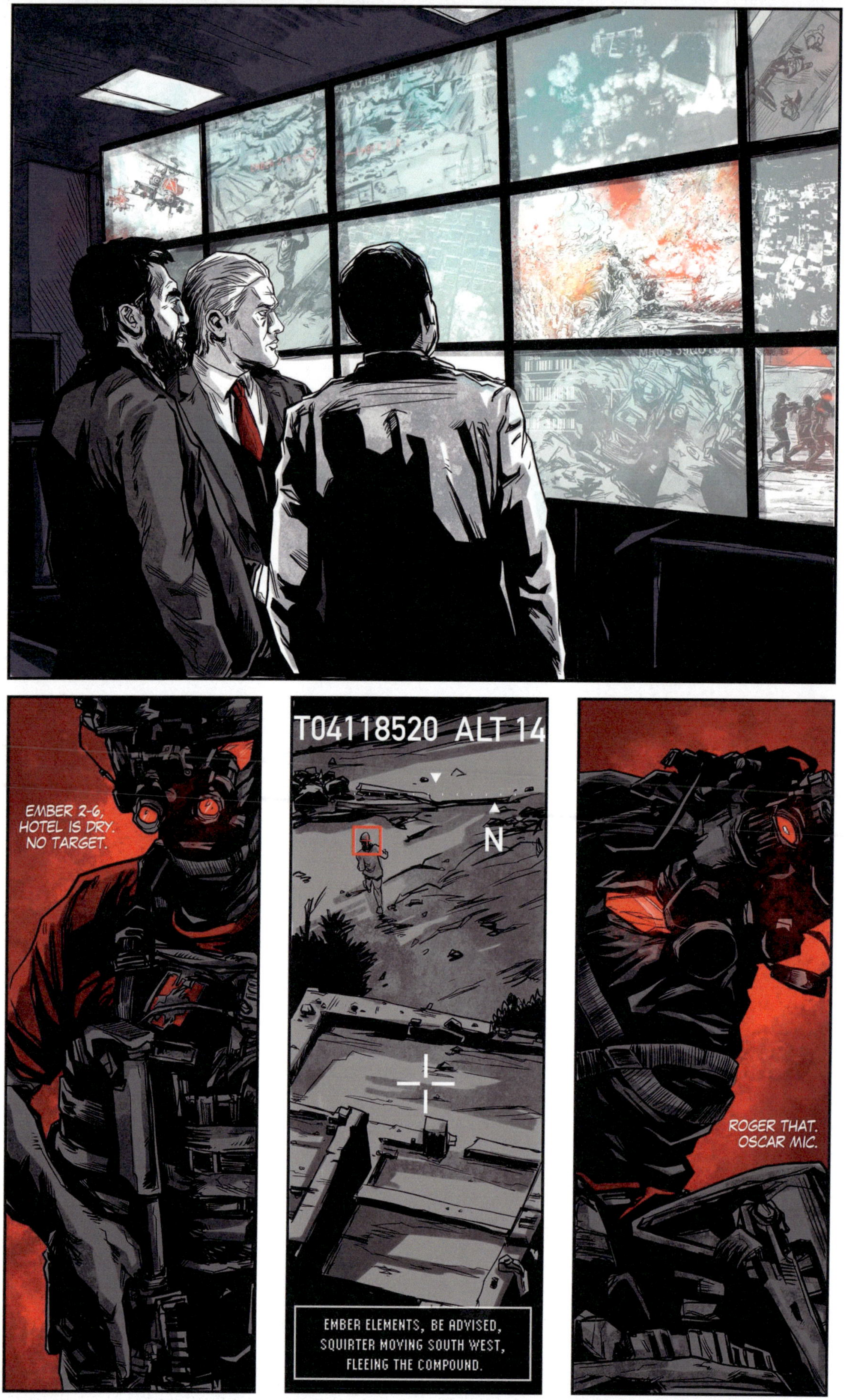
EMBER 2-6, HOTEL IS DRY. NO TARGET.
T04118520 ALT 14
N
EMBER ELEMENTS, BE ADVISED, SQUIRTER MOVING SOUTH WEST, FLEEING THE COMPOUND.
ROGER THAT. OSCAR MIC.

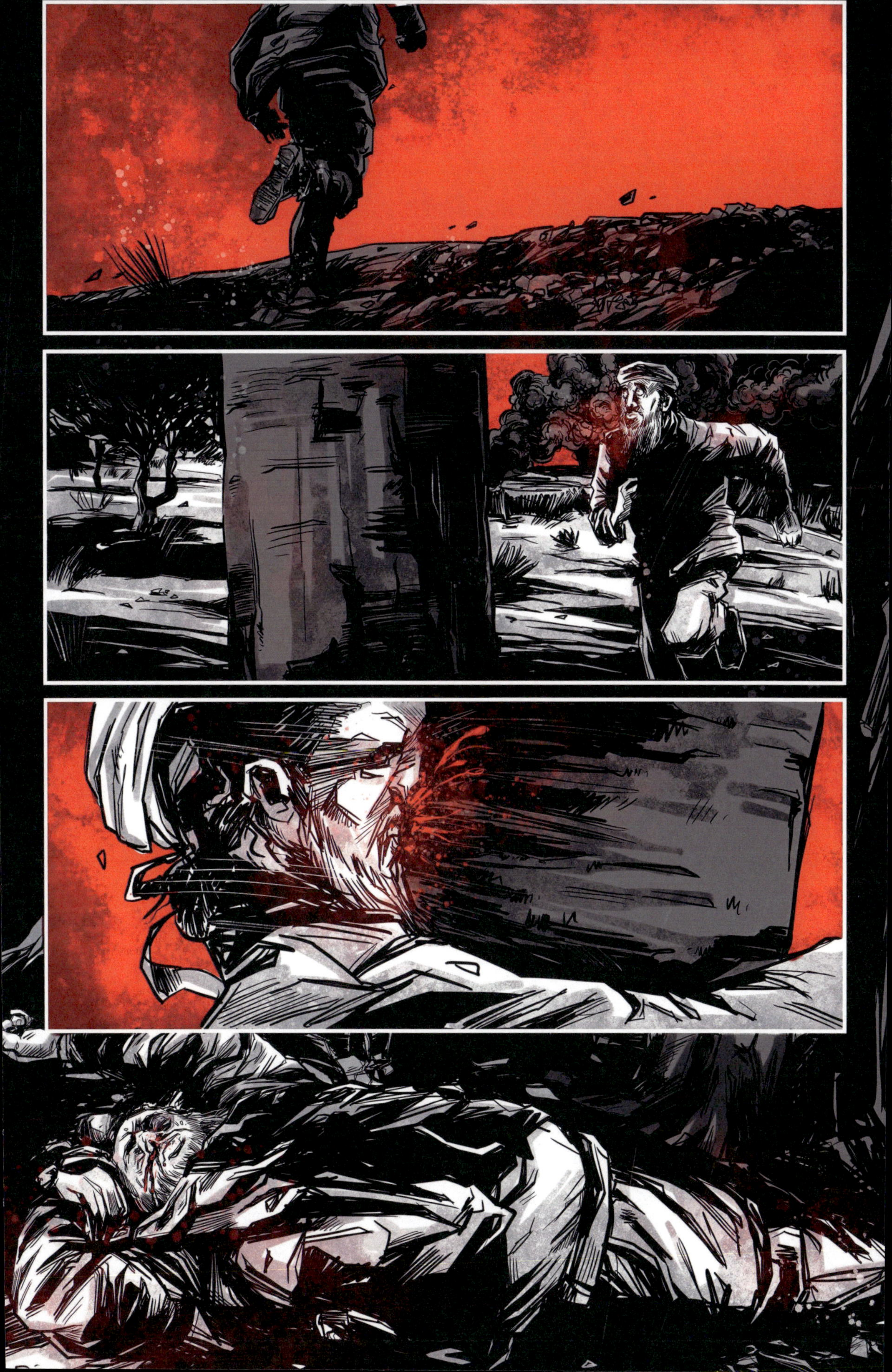

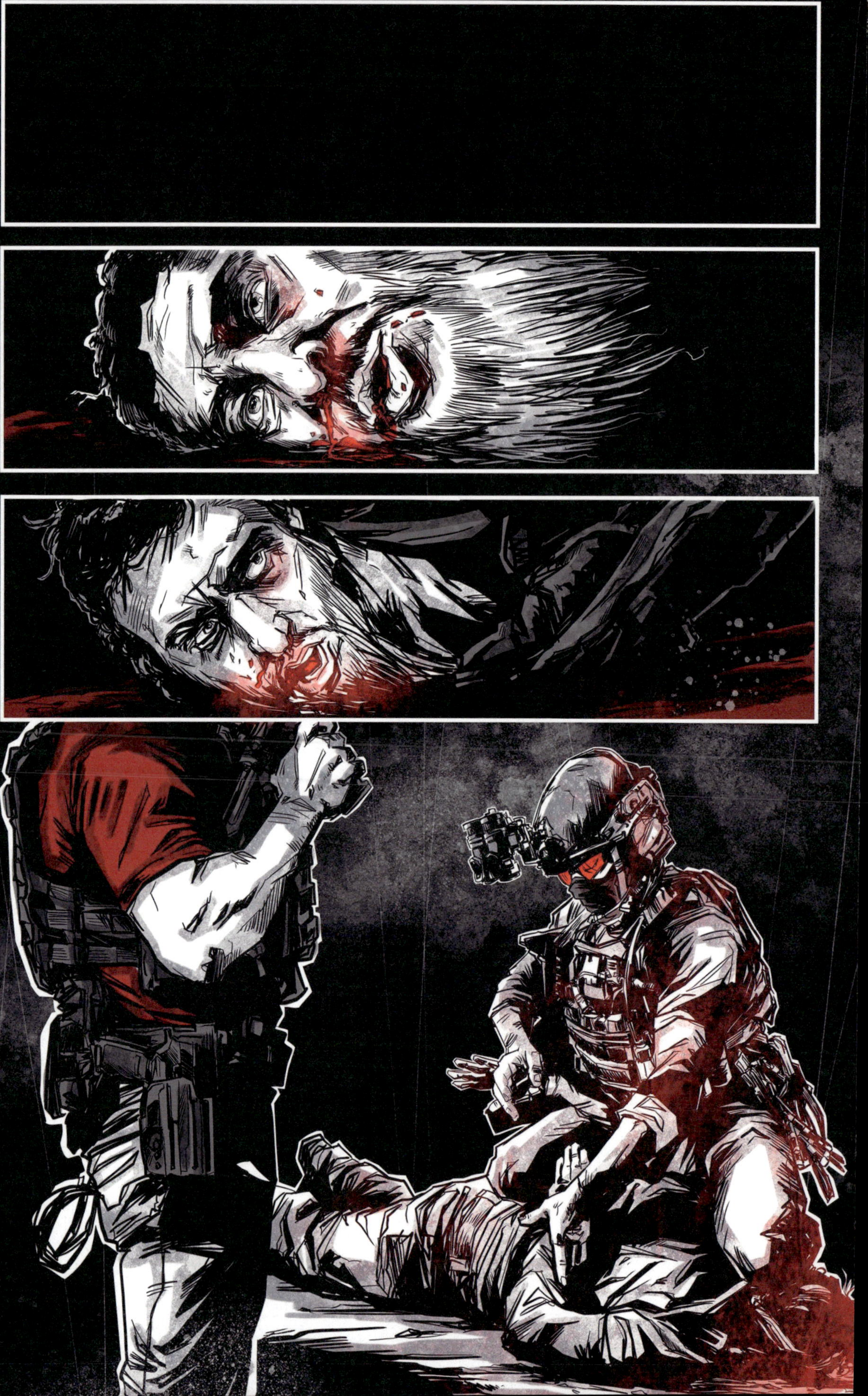

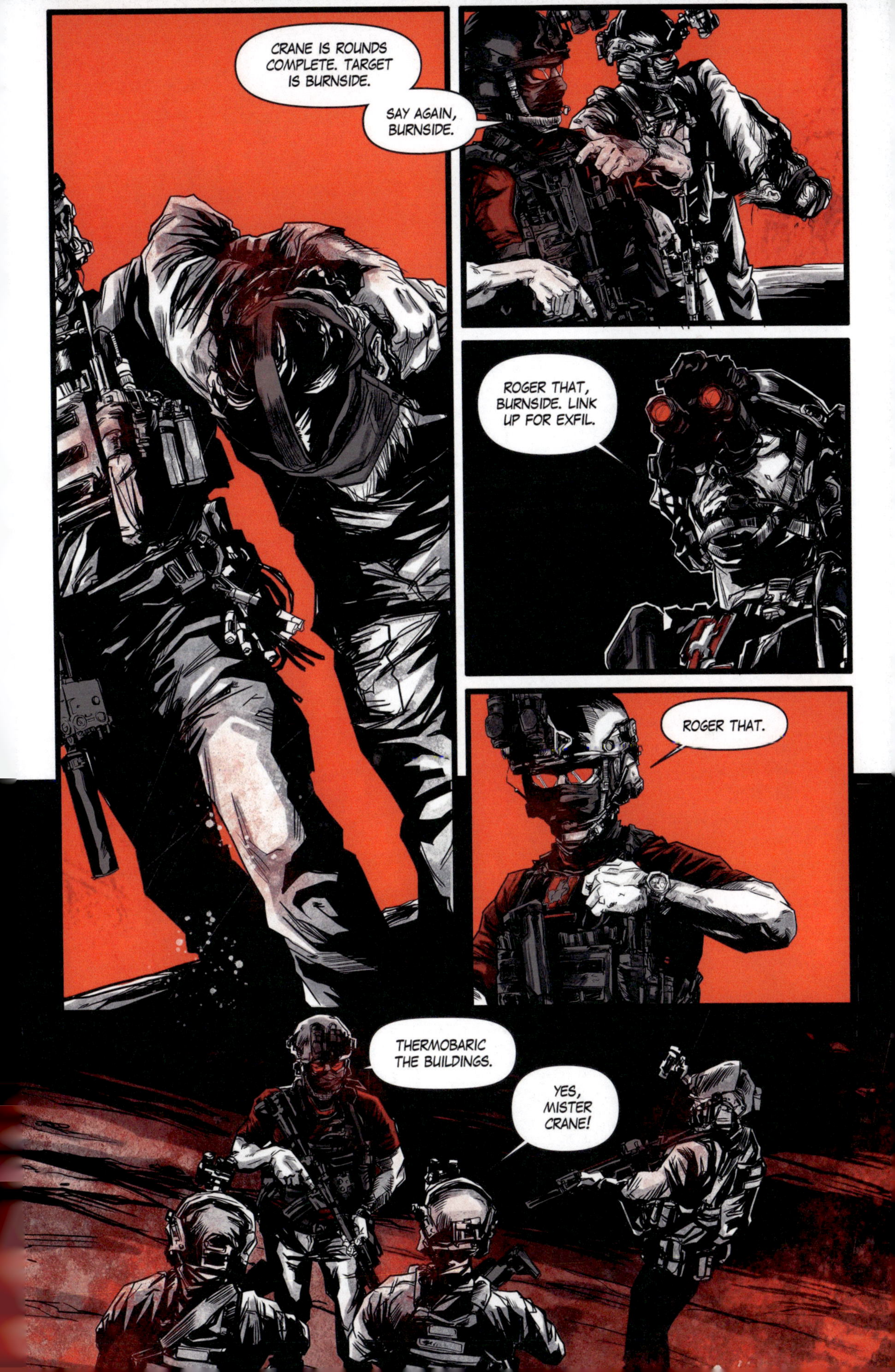
CRANE IS ROUNDS COMPLETE. TARGET IS BURNSIDE.
SAY AGAIN, BURNSIDE.
ROGER THAT, BURNSIDE. LINK UP FOR EXFIL.
ROGER THAT.
THERMOBARIC THE BUILDINGS.
YES, MISTER CRANE!

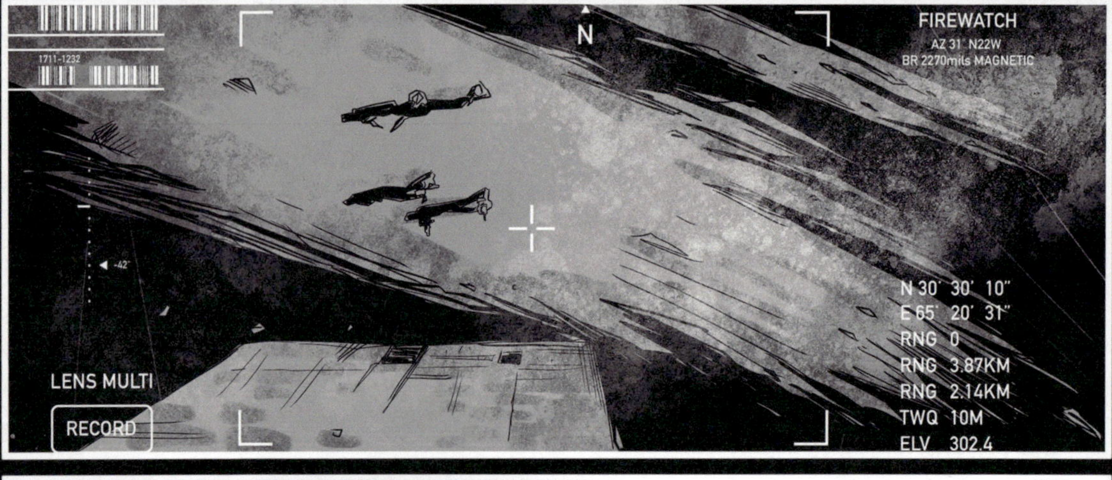
1711-1232
N
FIREWATCH
AZ 31° N22W
BR 2270mils MAGNETIC
-42°
N 30' 30' 10"
E 65' 20' 31"
RNG 0
RNG 3.87KM
RNG 2.14KM
TWQ 10M
ELV 302.4
LENS MULTI
RECORD

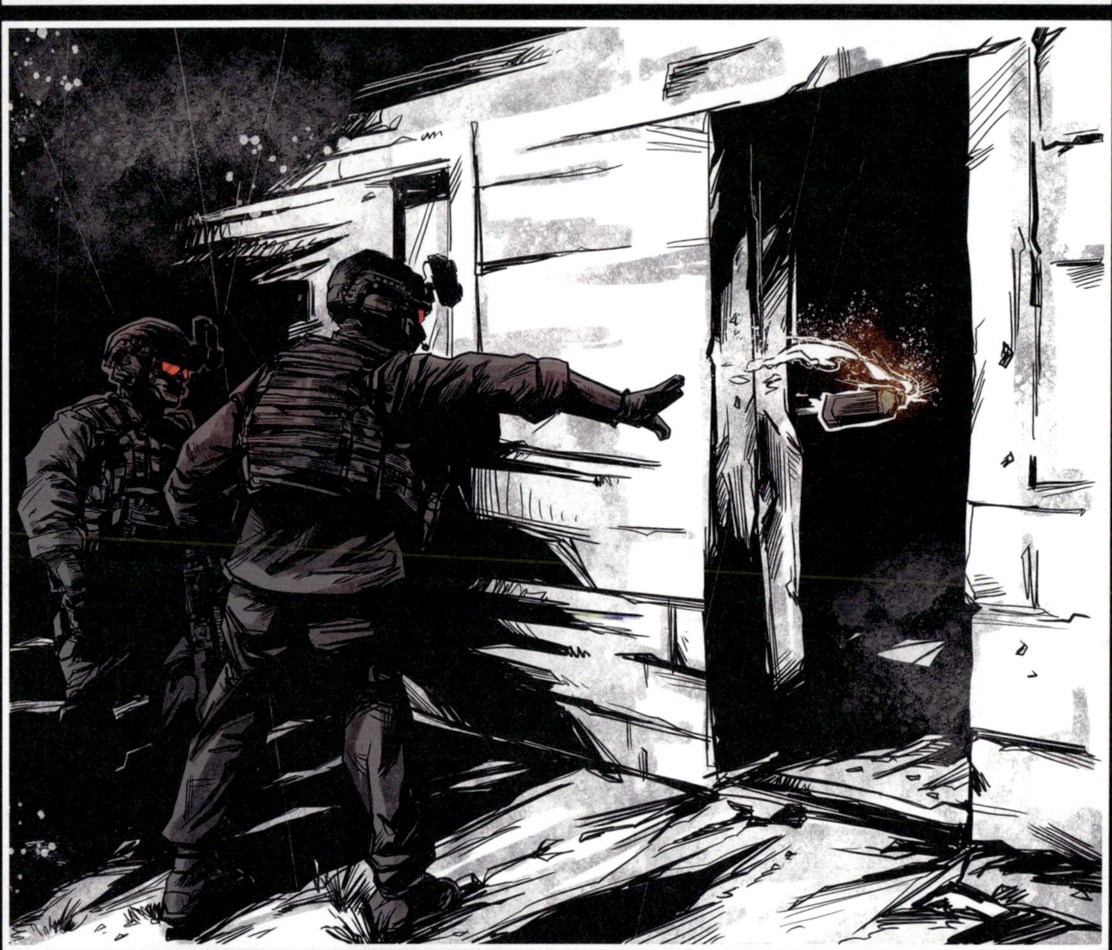

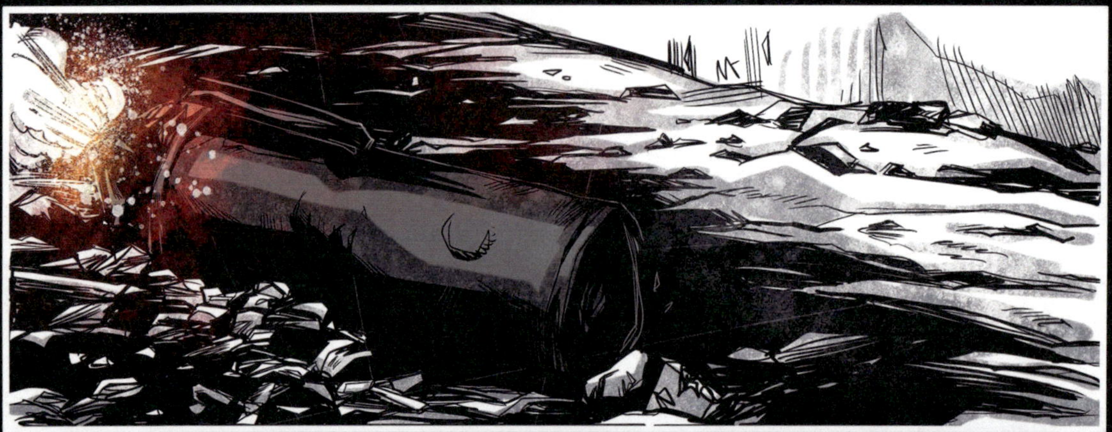

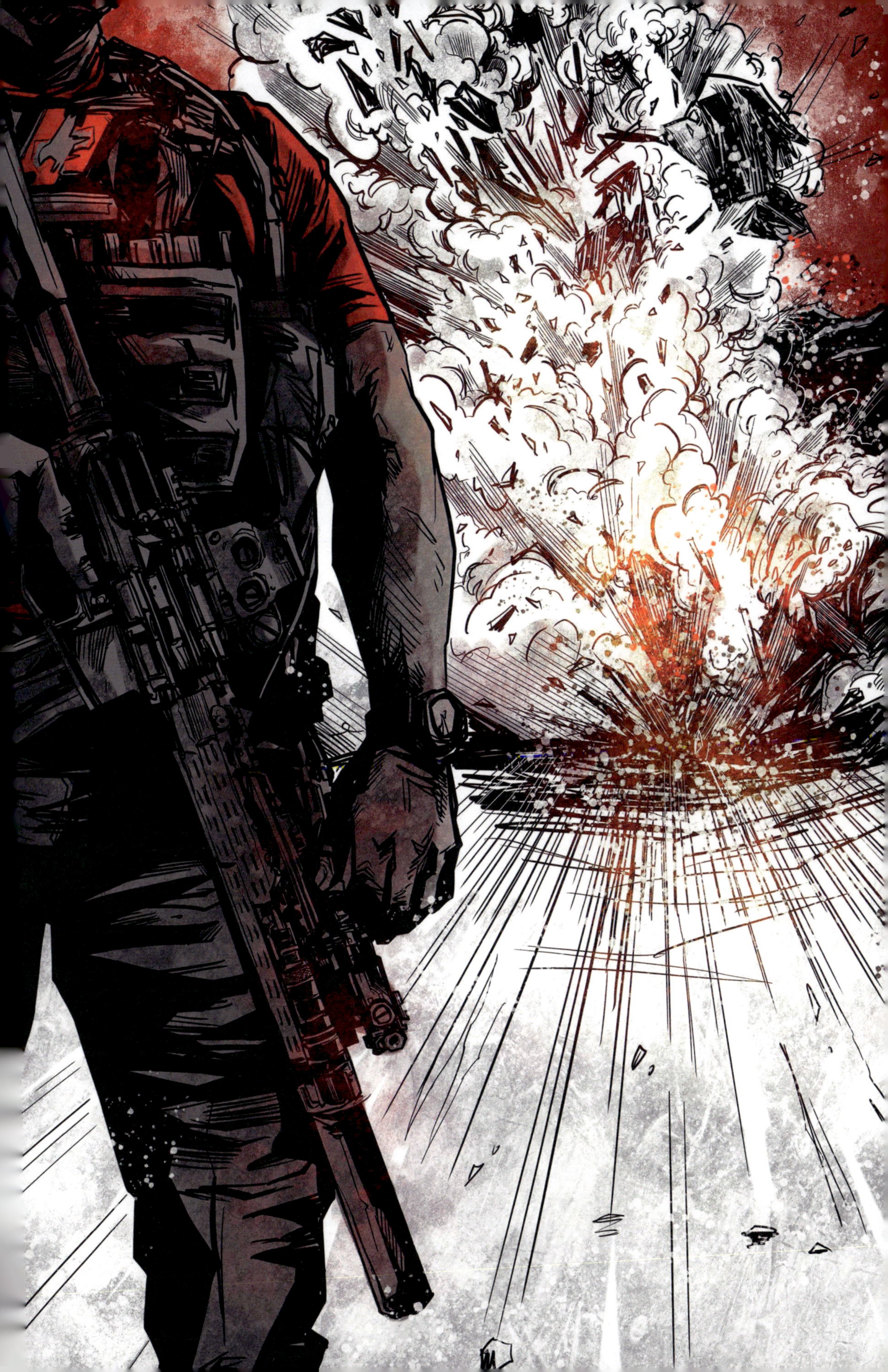

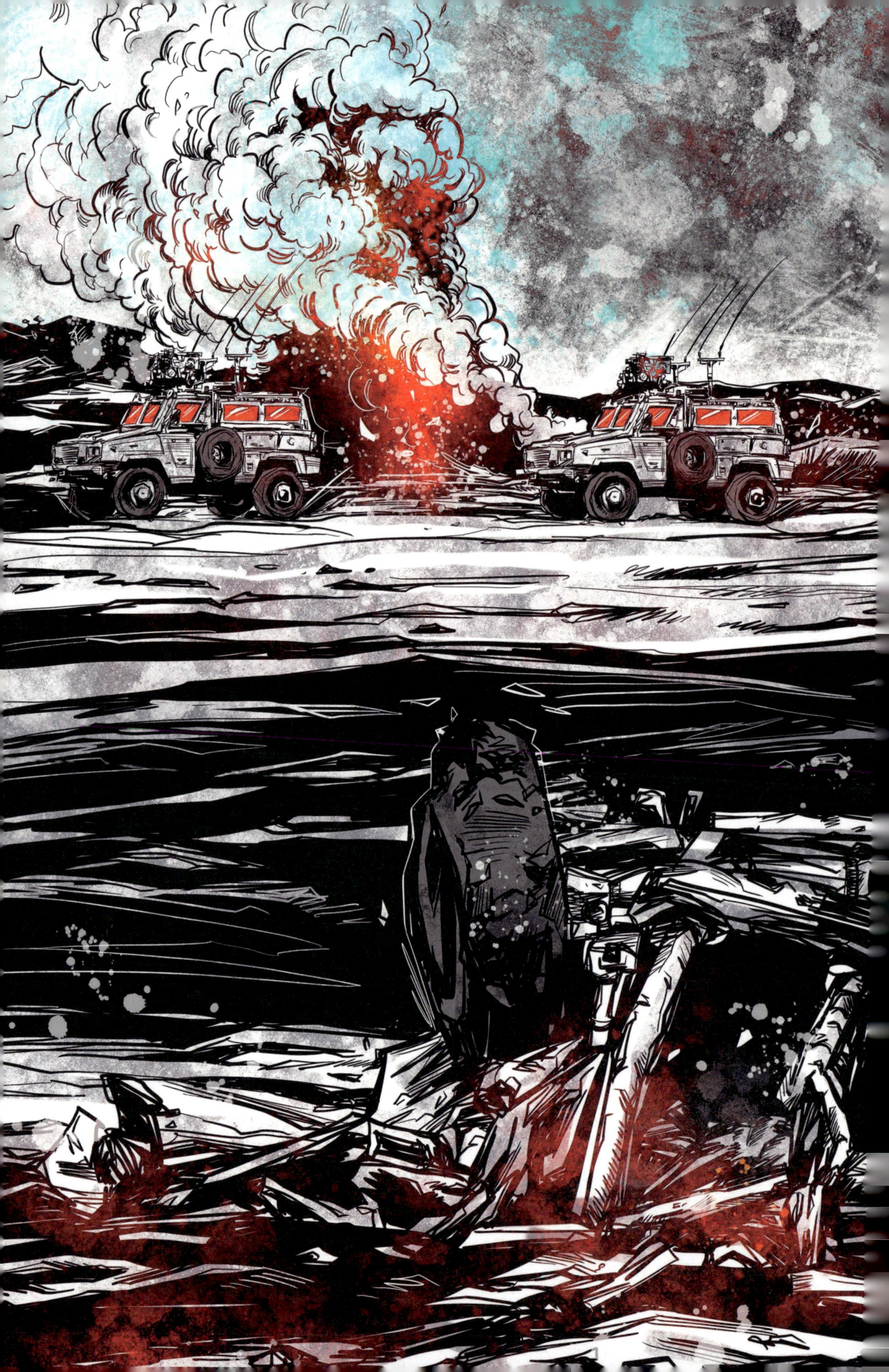

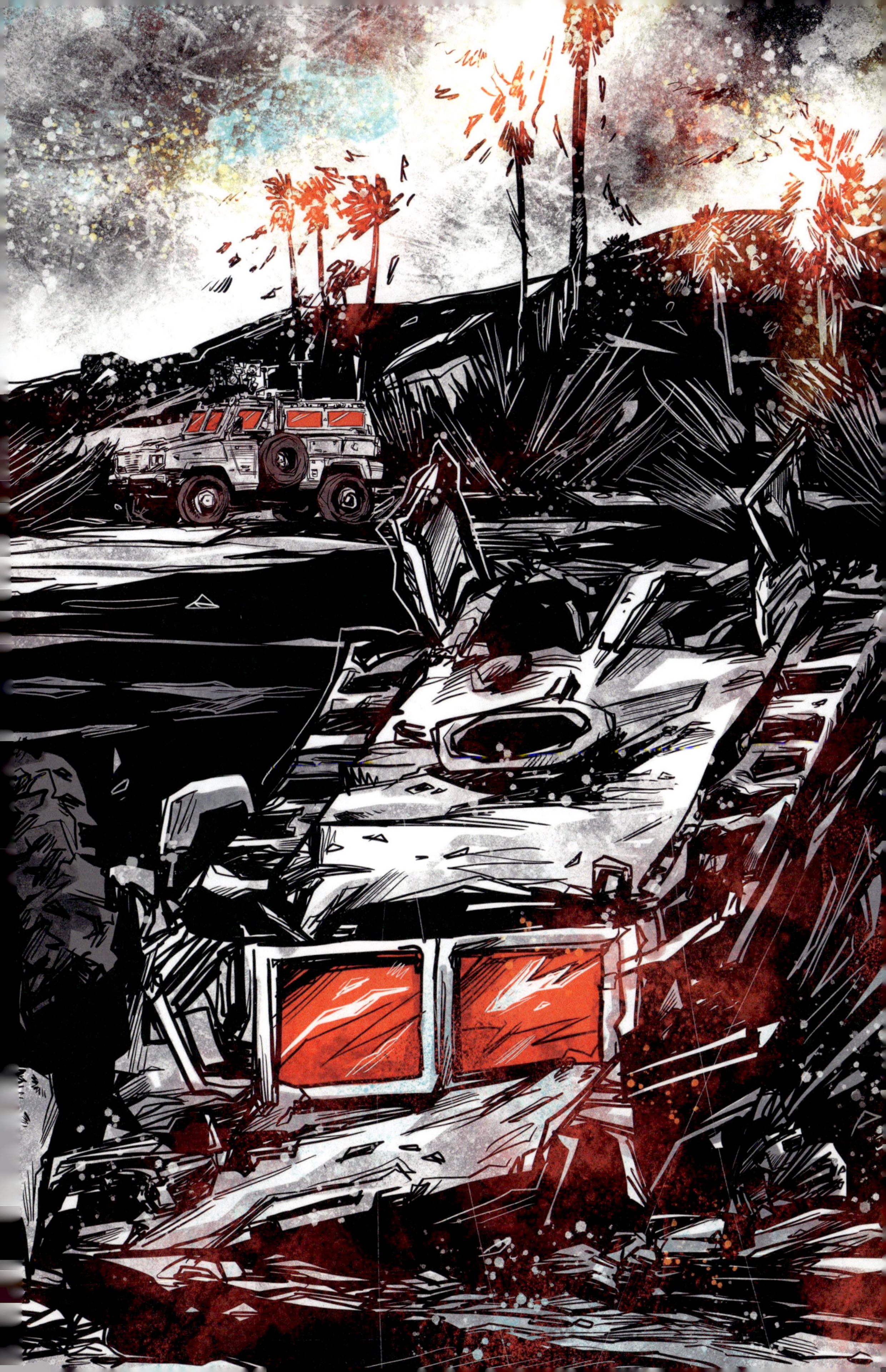

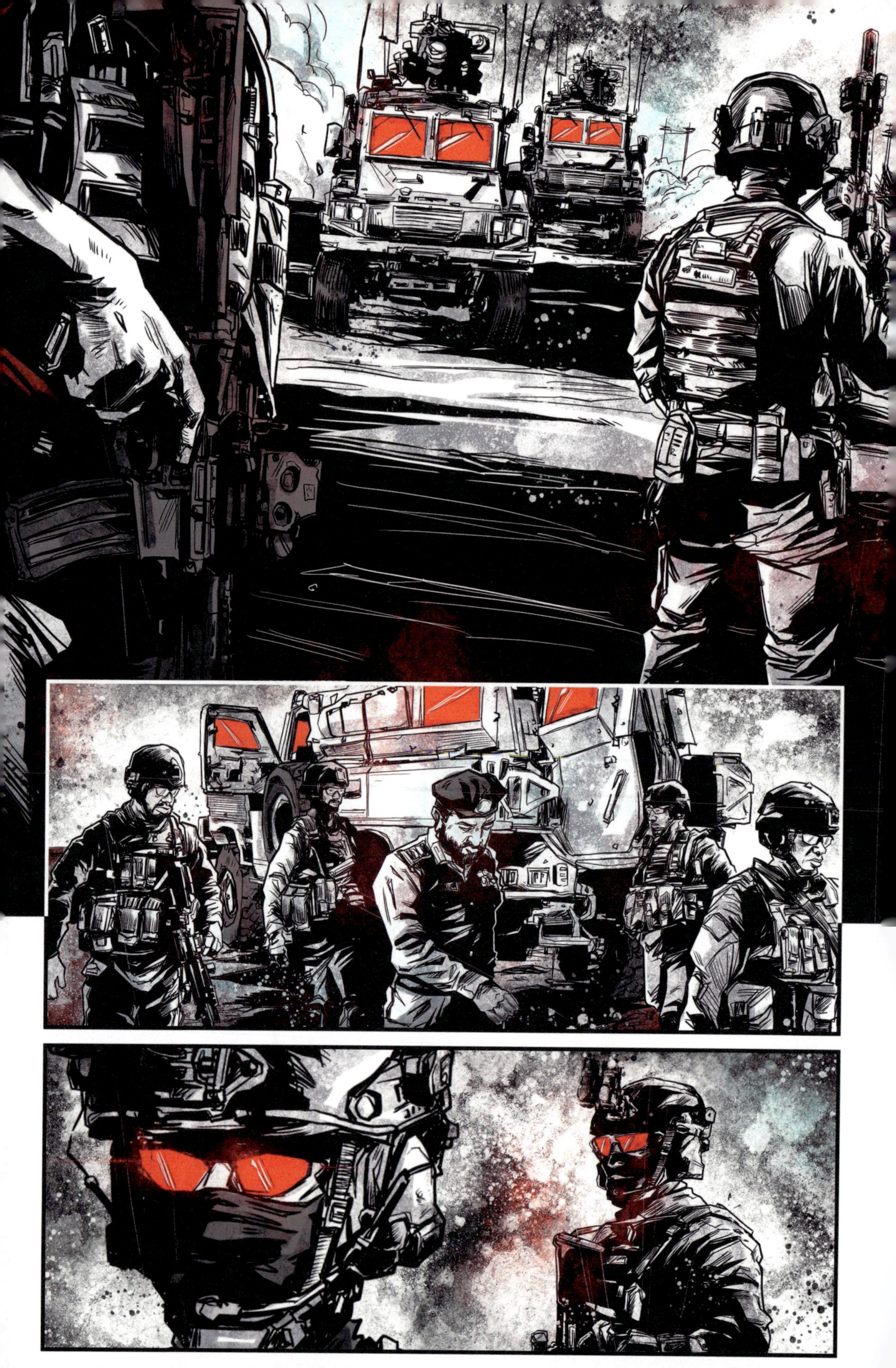

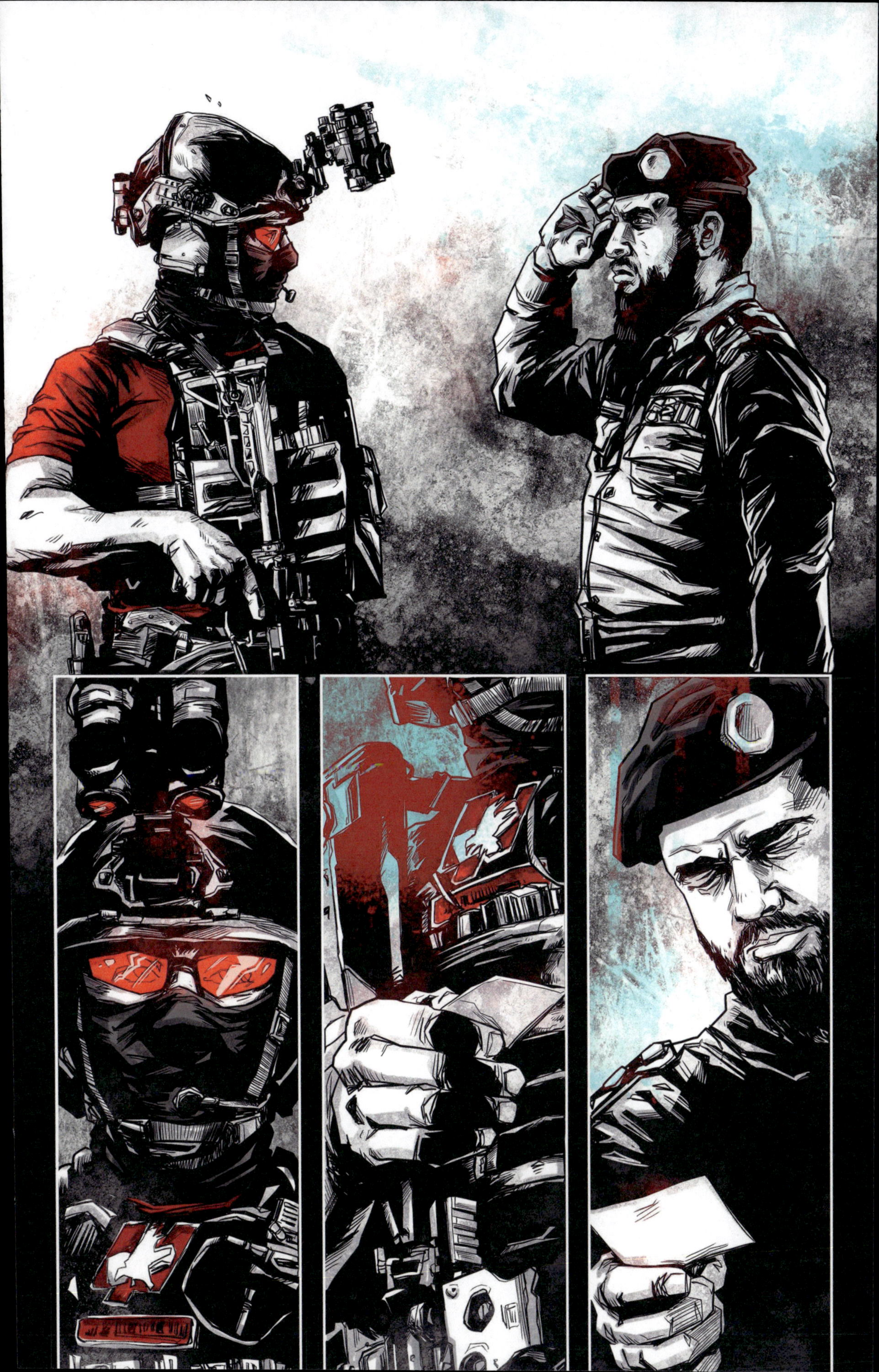

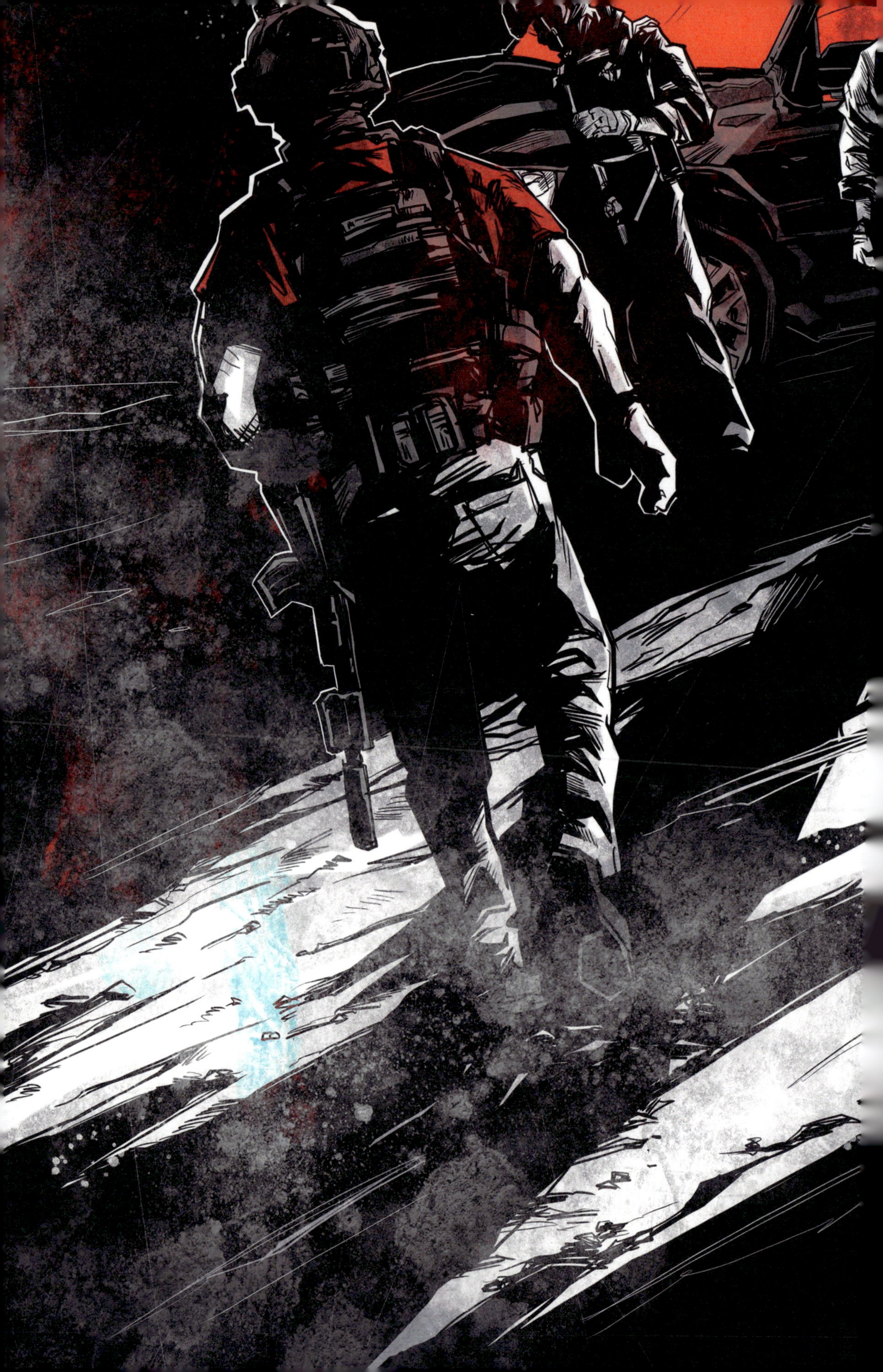

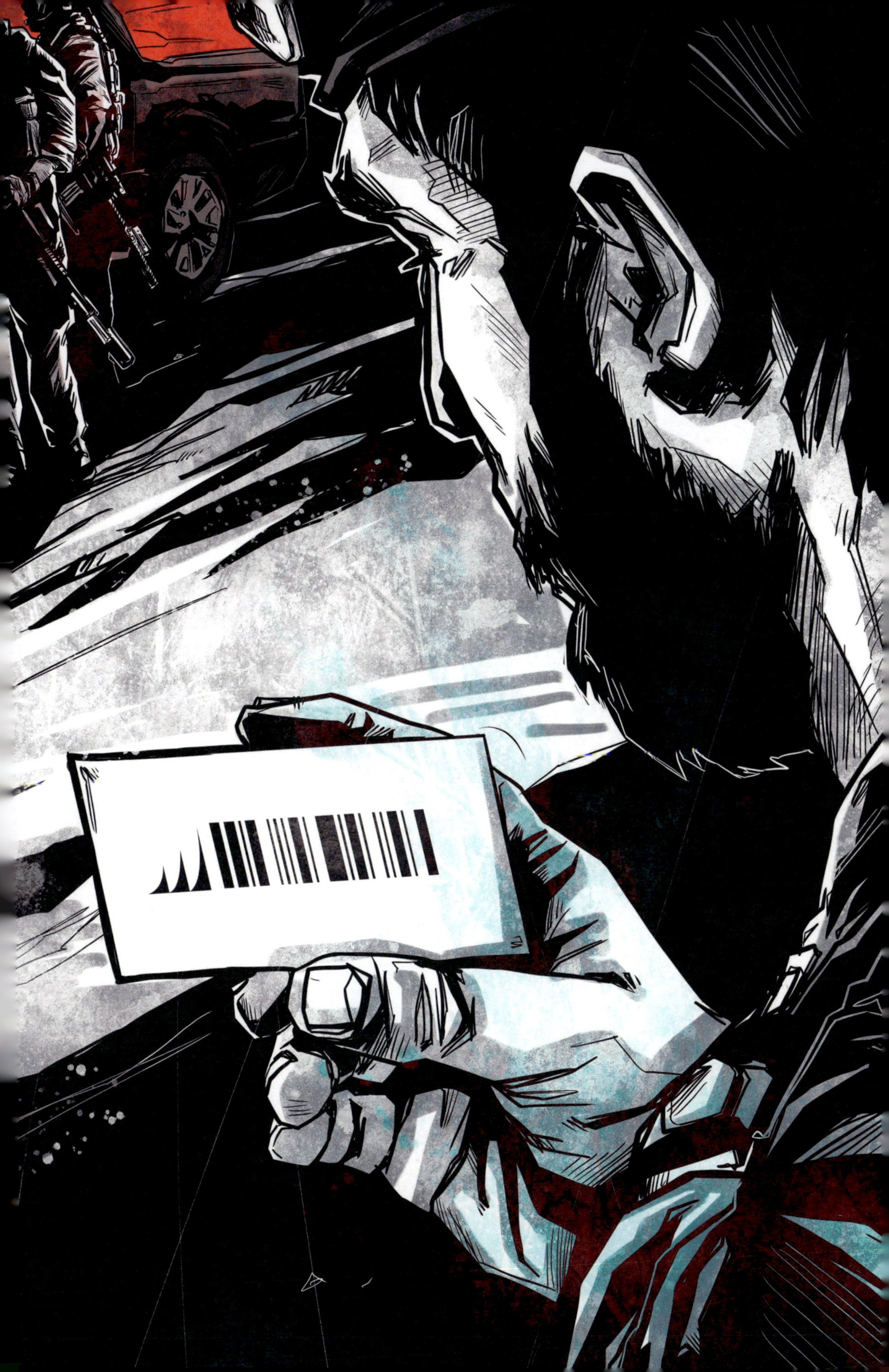

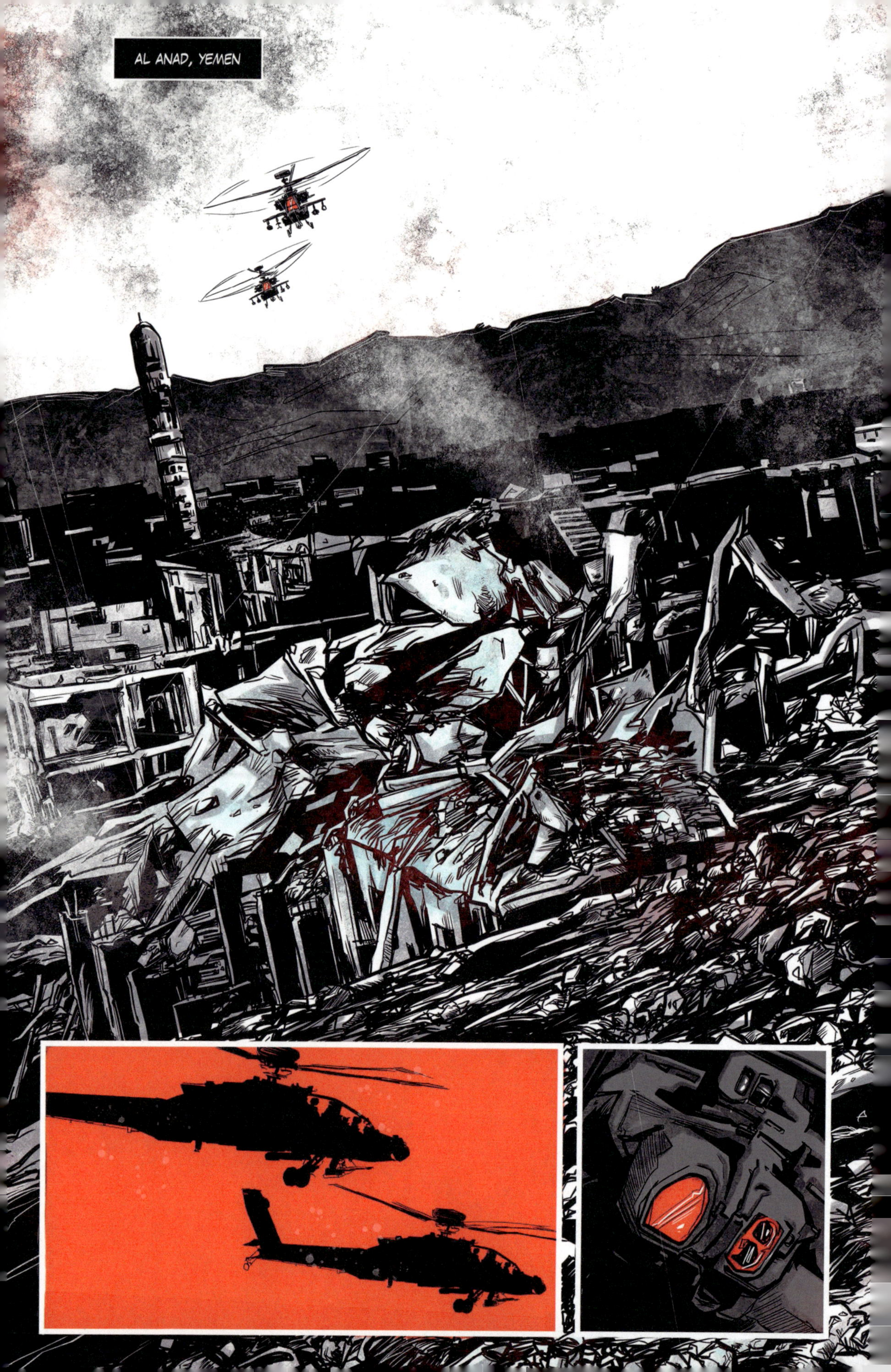
AL ANAD, YEMEN

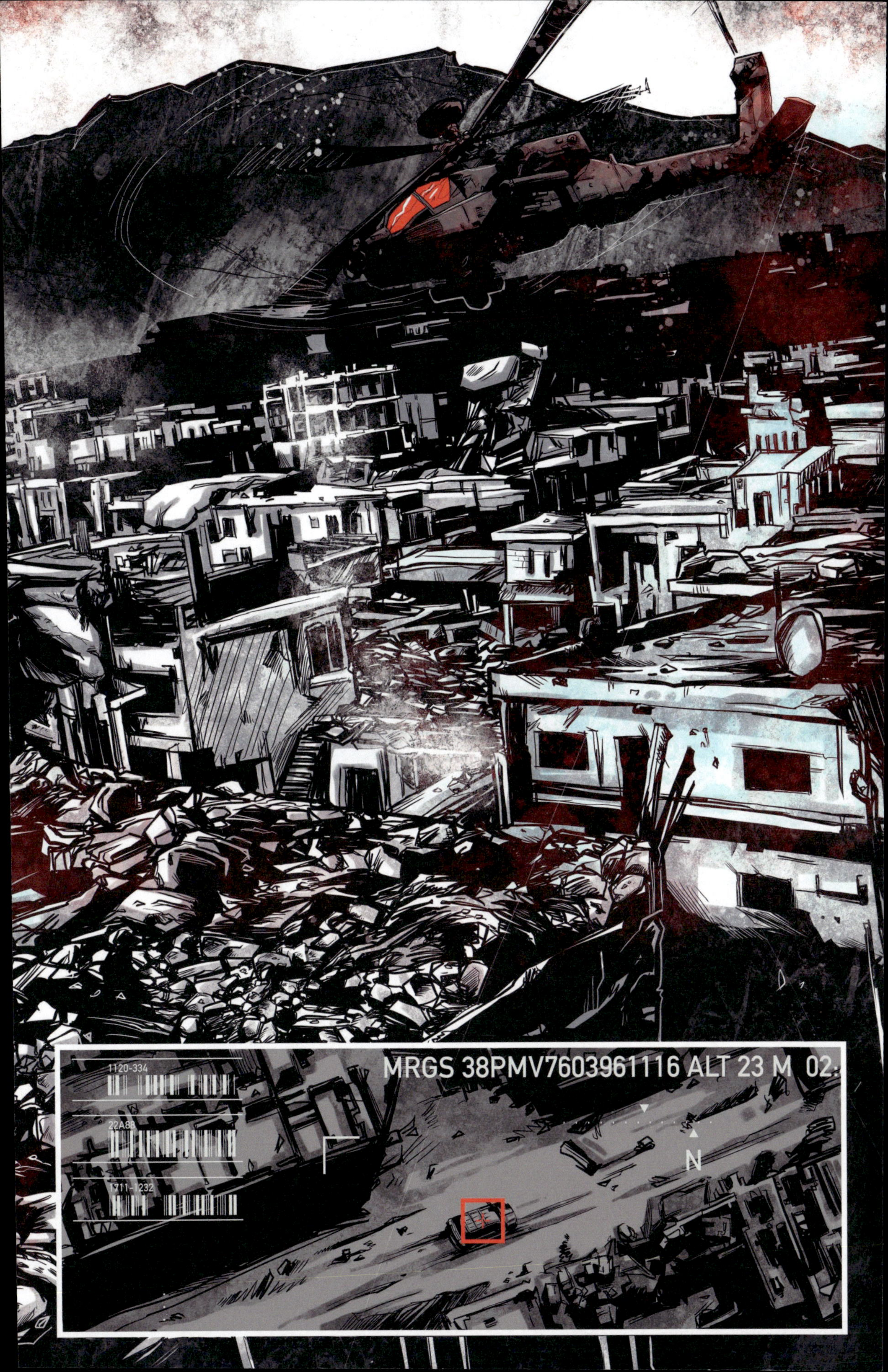
MRGS 38PMV7603961116 ALT 23 M 02:
1120-334
22A88
1711-1232
N

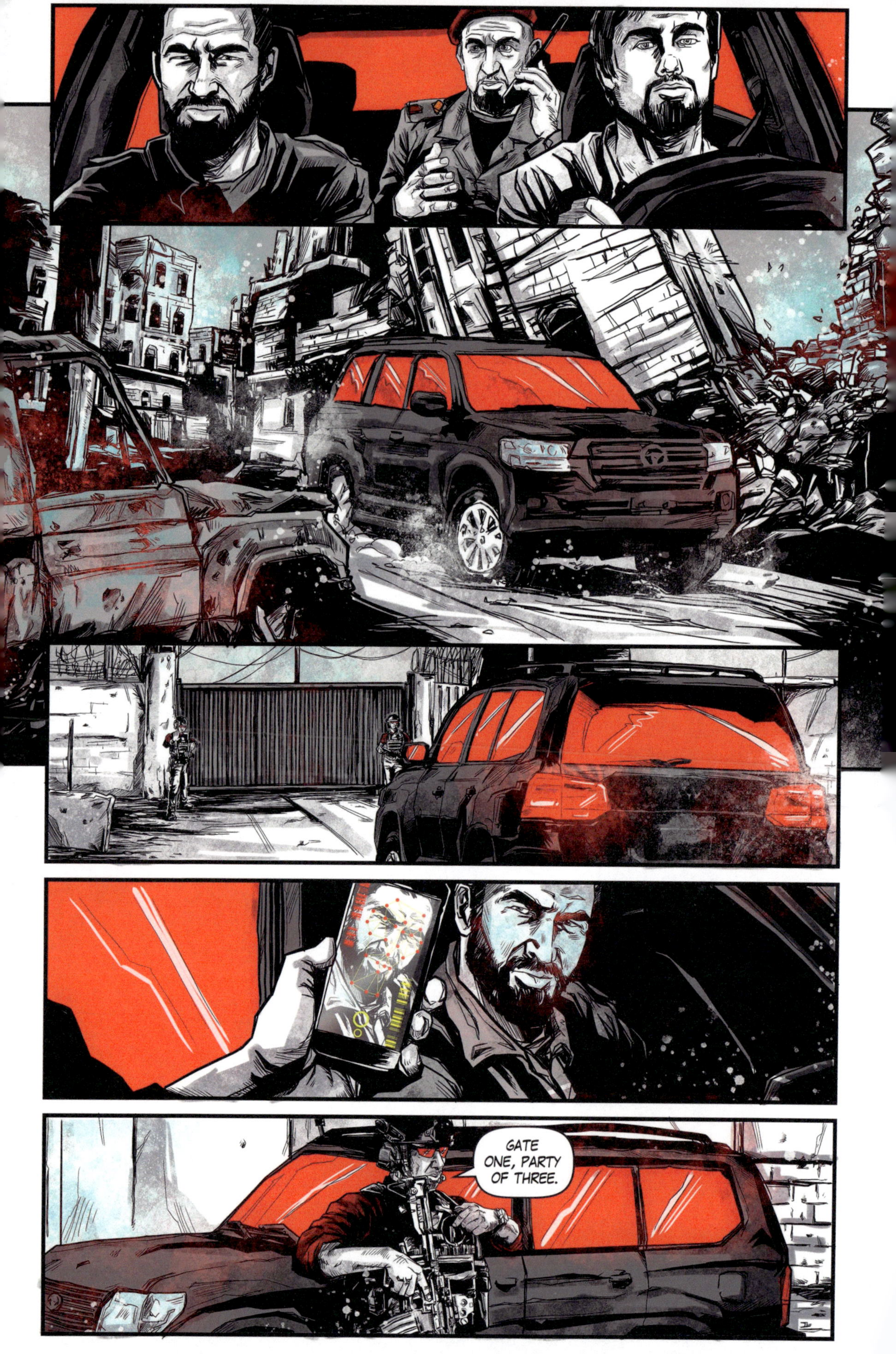
GATE ONE, PARTY OF THREE.

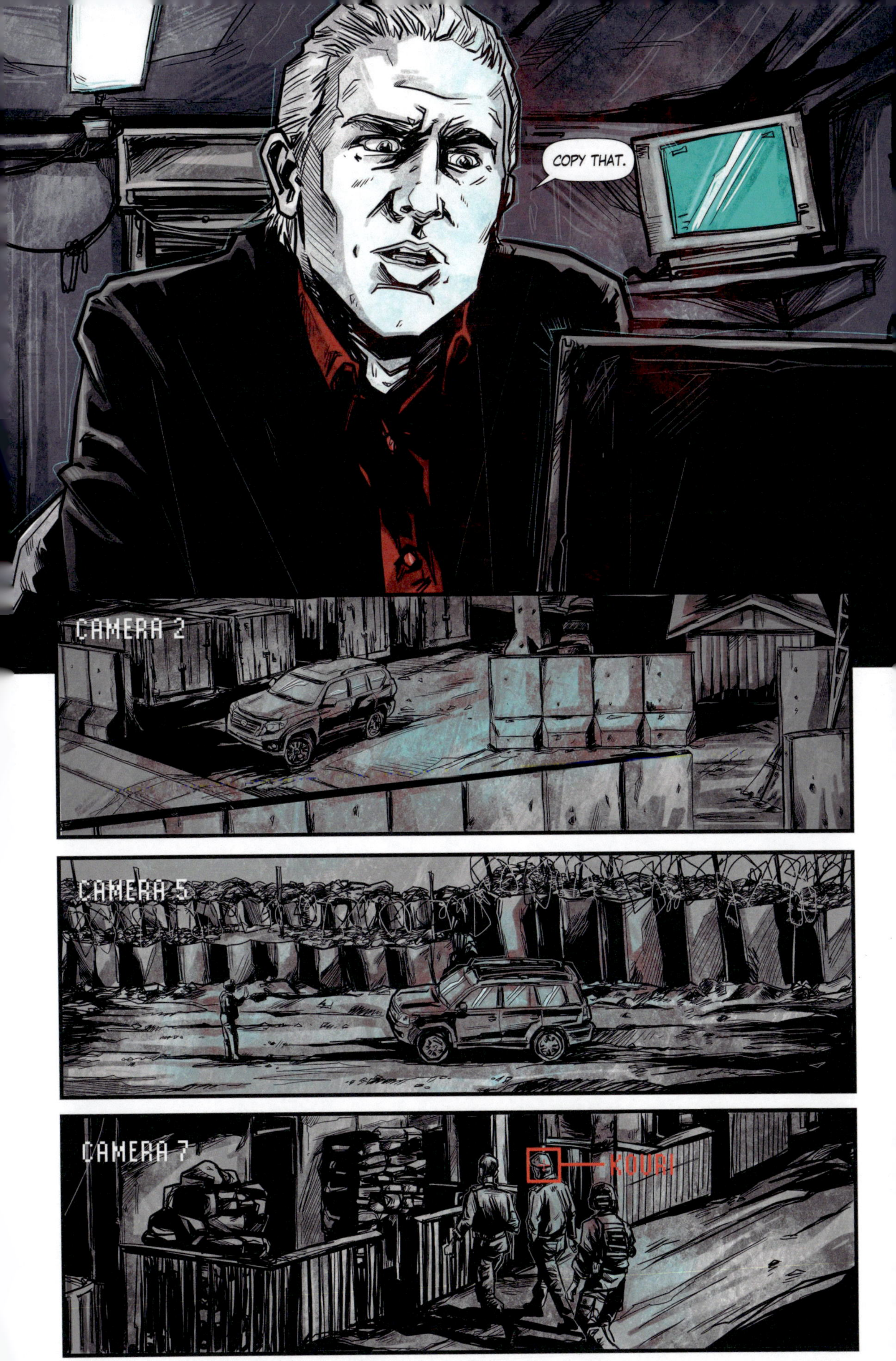
COPY THAT.
CAMERA 2
CAMERA 5
CAMERA 7
KOURI

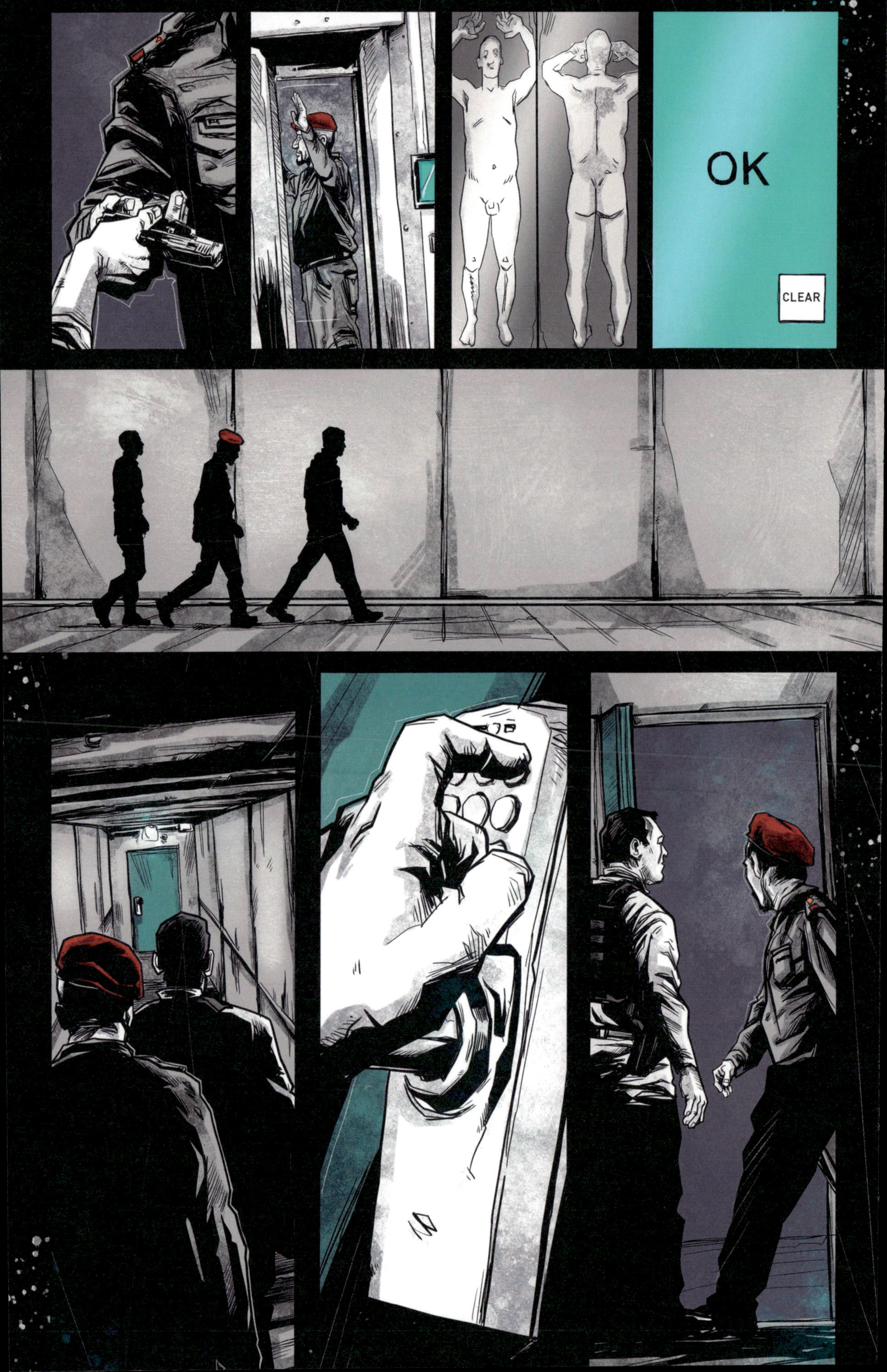
OK
CLEAR

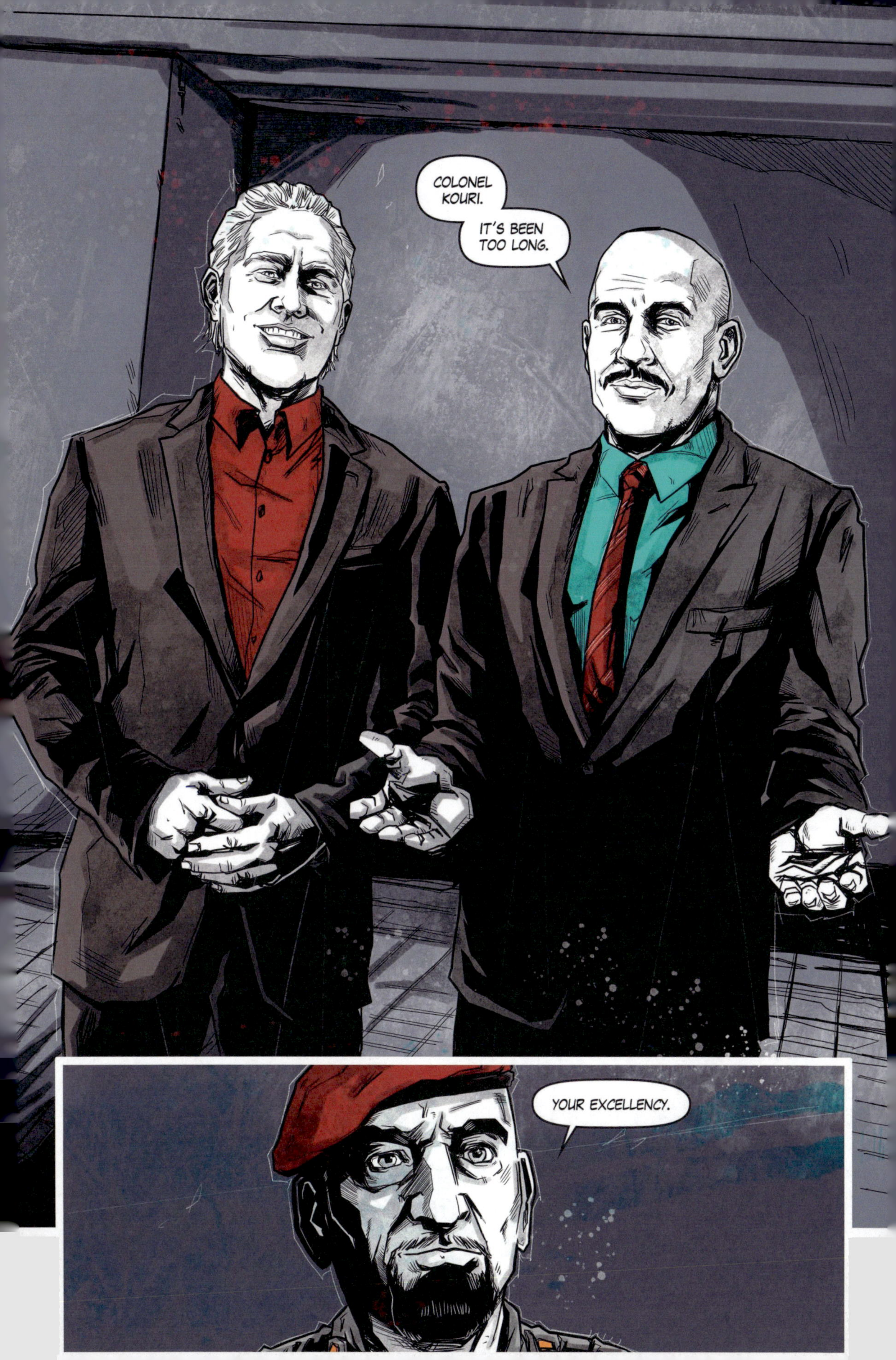
COLONEL KOURI.
IT'S BEEN TOO LONG.
YOUR EXCELLENCY.

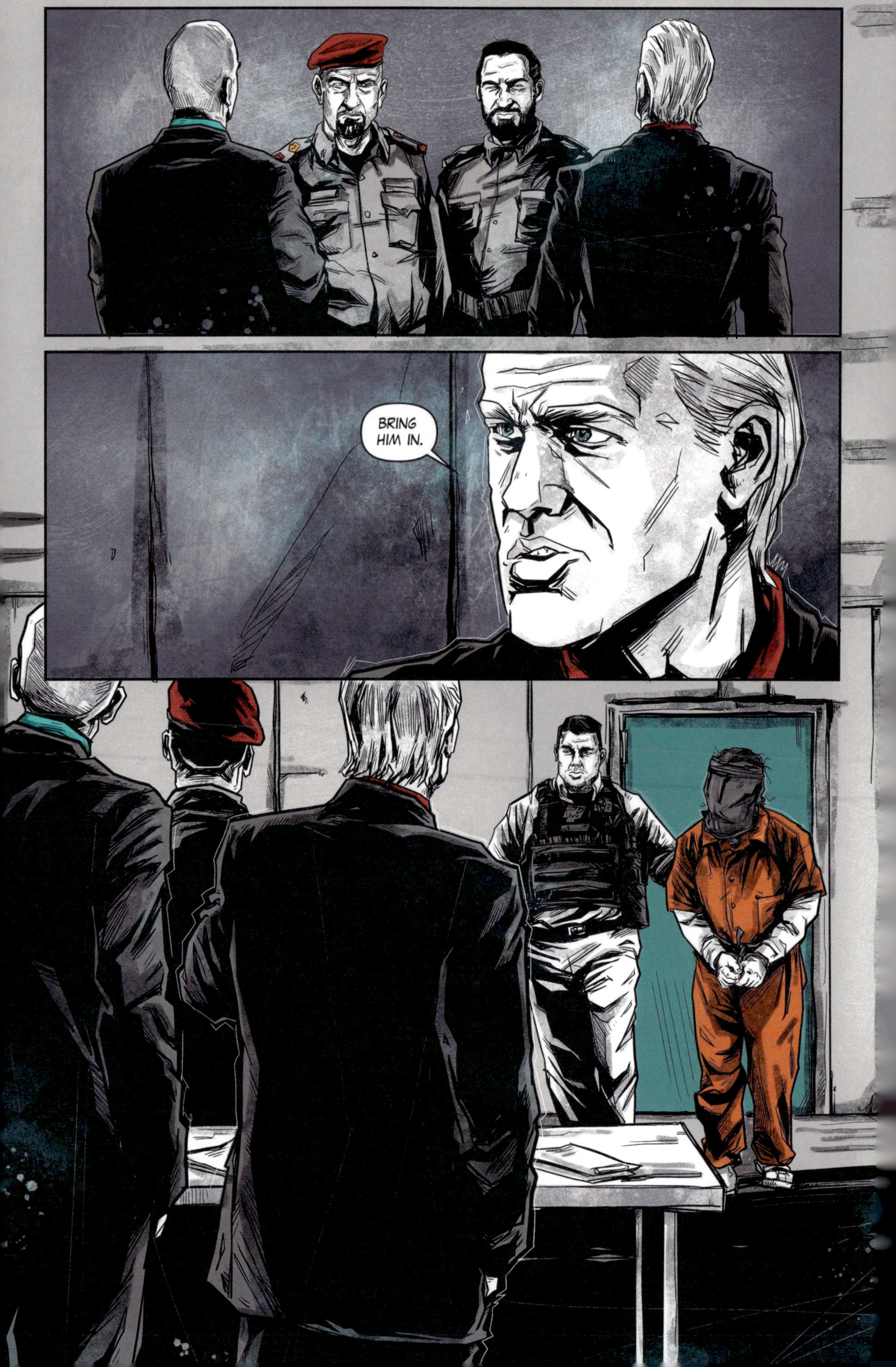
BRING HIM IN.

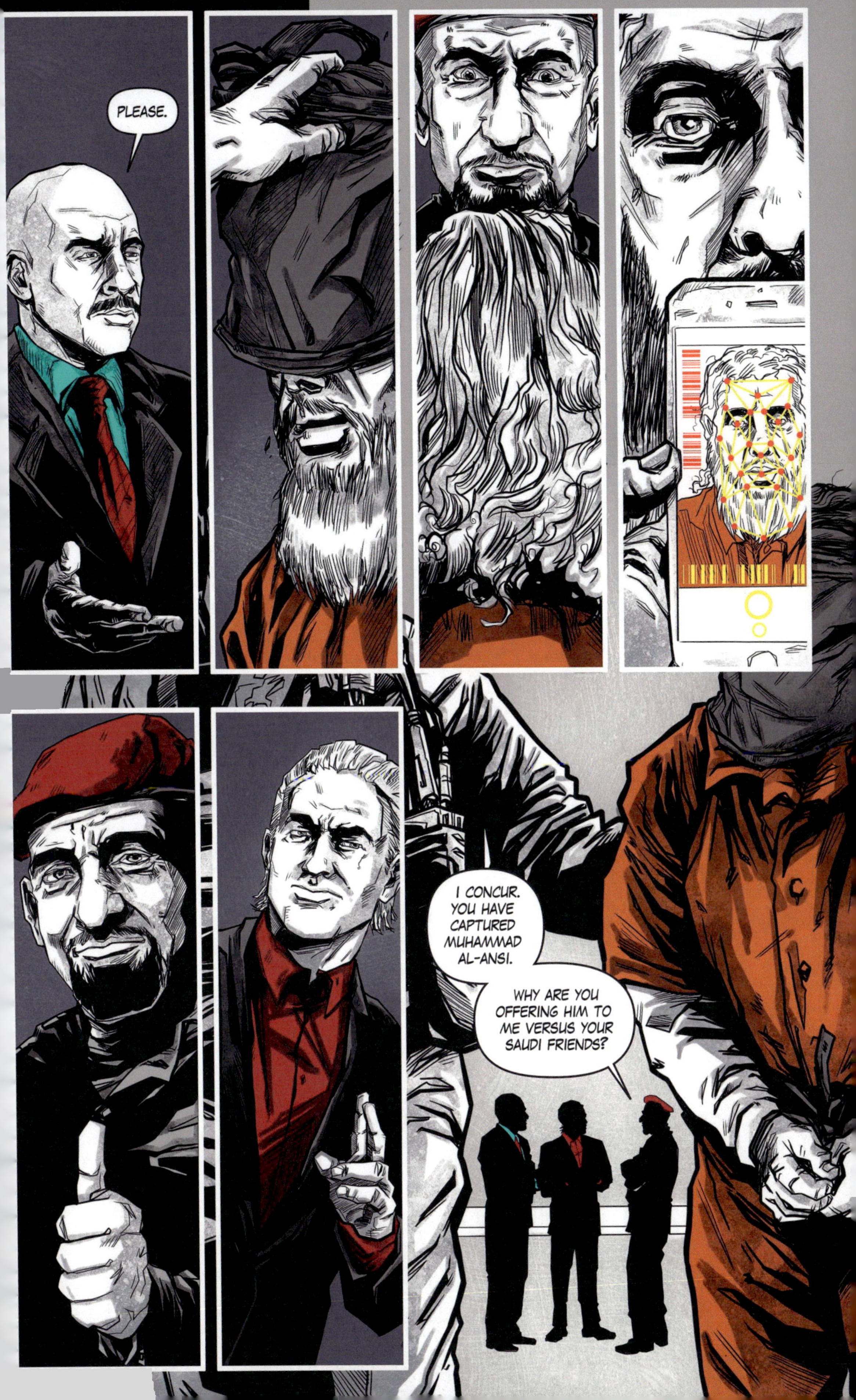
PLEASE.
I CONCUR. YOU HAVE CAPTURED MUHAMMAD AL-ANSI.
WHY ARE YOU OFFERING HIM TO ME VERSUS YOUR SAUDI FRIENDS?

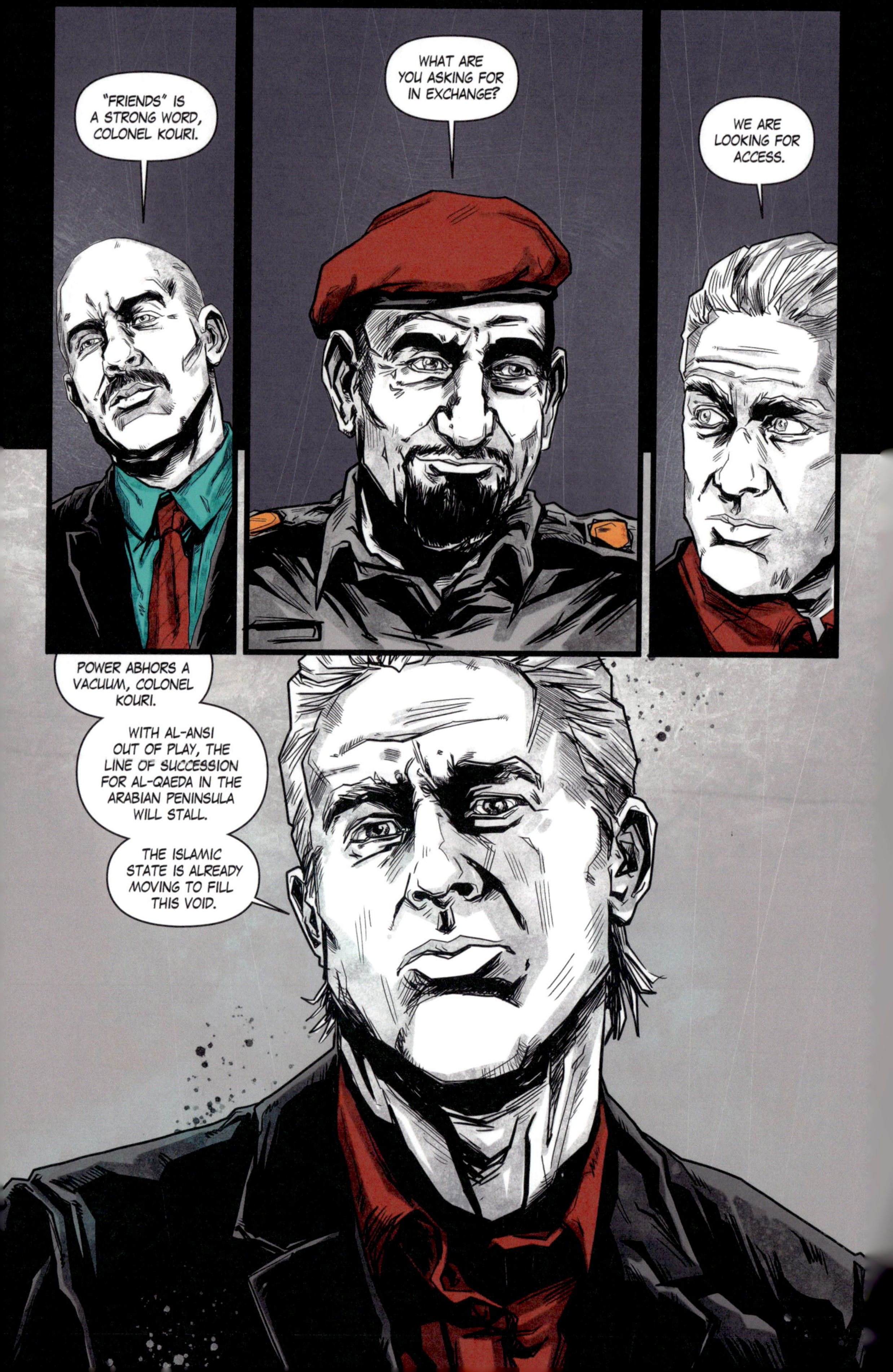
"FRIENDS" IS A STRONG WORD, COLONEL KOURI.
WHAT ARE YOU ASKING FOR IN EXCHANGE?
WE ARE LOOKING FOR ACCESS.
POWER ABHORS A VACUUM, COLONEL KOURI.
WITH AL-ANSI OUT OF PLAY, THE LINE OF SUCCESSION FOR AL-QAEDA IN THE ARABIAN PENINSULA WILL STALL.
THE ISLAMIC STATE IS ALREADY MOVING TO FILL THIS VOID.

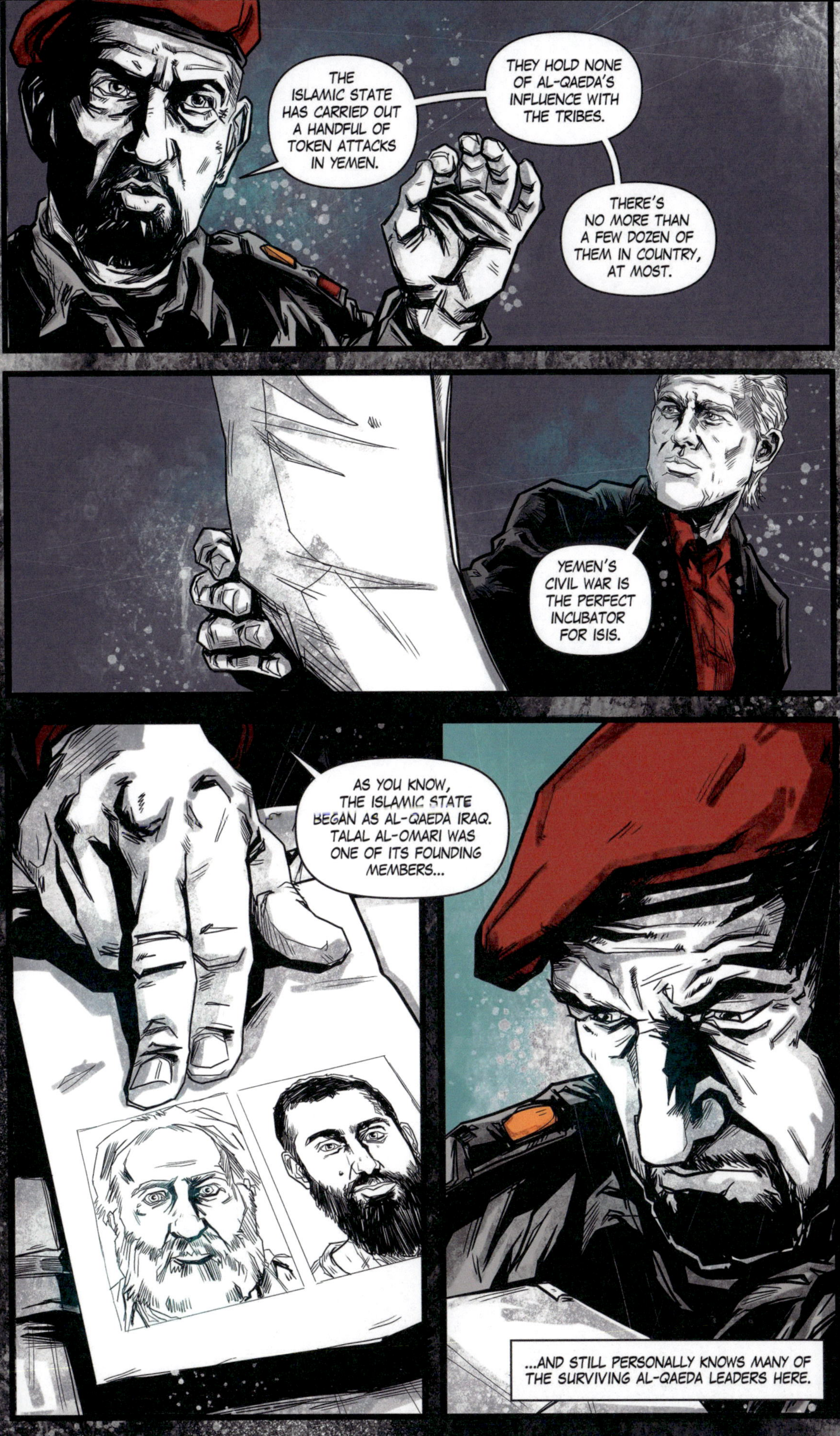
THE ISLAMIC STATE HAS CARRIED OUT A HANDFUL OF TOKEN ATTACKS IN YEMEN.
THEY HOLD NONE OF AL-QAEDA'S INFLUENCE WITH THE TRIBES.
THERE'S NO MORE THAN A FEW DOZEN OF THEM IN COUNTRY, AT MOST.
YEMEN'S CIVIL WAR IS THE PERFECT INCUBATOR FOR ISIS.
AS YOU KNOW, THE ISLAMIC STATE BEGAN AS AL-QAEDA IRAQ. TALAL AL-OMARI WAS ONE OF ITS FOUNDING MEMBERS...
...AND STILL PERSONALLY KNOWS MANY OF THE SURVIVING AL-QAEDA LEADERS HERE.

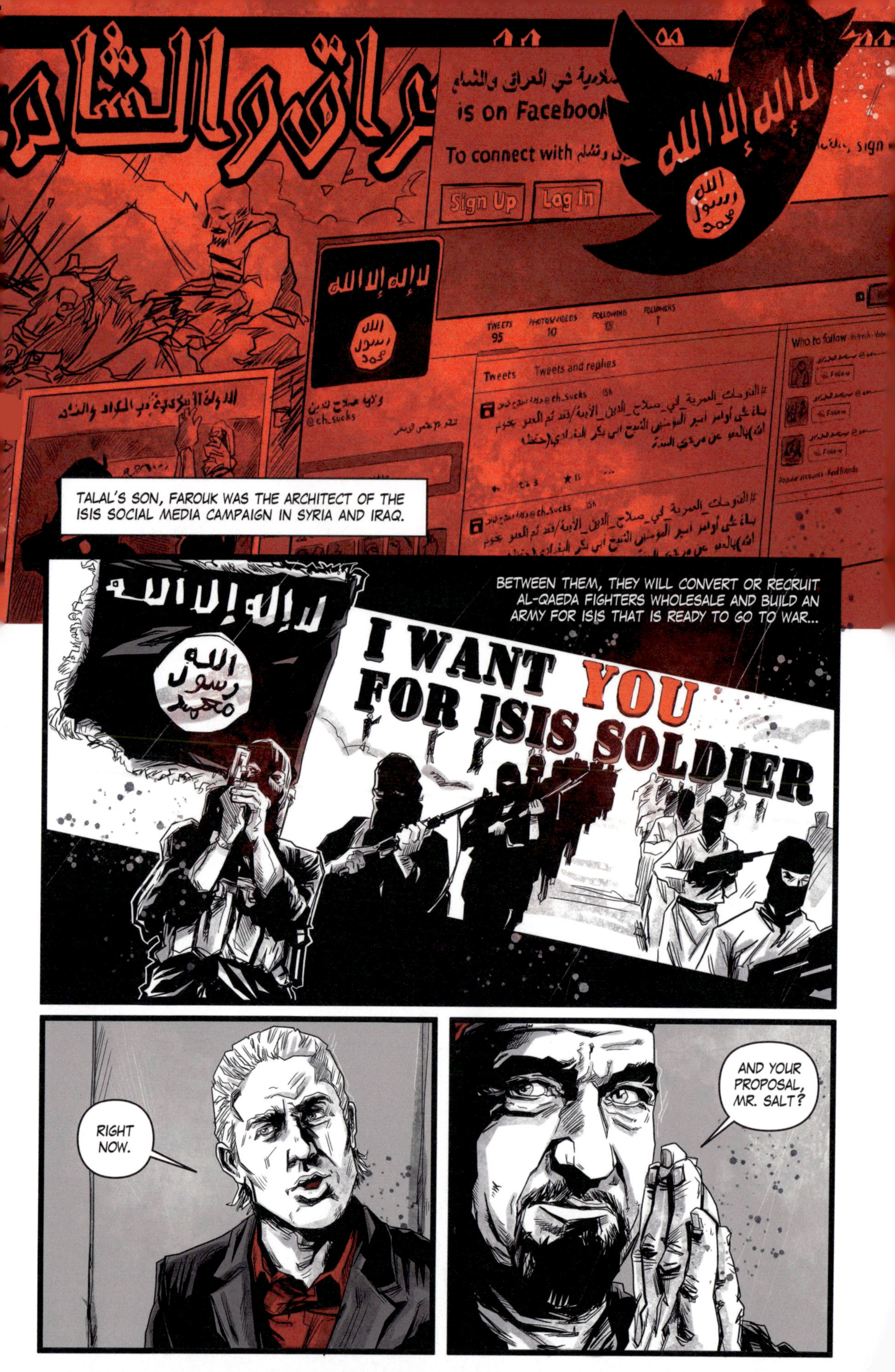
Sign Up
Log In
TALAL'S SON, FAROUK WAS THE ARCHITECT OF THE ISIS SOCIAL MEDIA CAMPAIGN IN SYRIA AND IRAQ.
BETWEEN THEM, THEY WILL CONVERT OR RECRUIT AL-QAEDA FIGHTERS WHOLESALE AND BUILD AN ARMY FOR ISIS THAT IS READY TO GO TO WAR...
I WANT YOU FOR ISIS SOLDIER
RIGHT NOW.
AND YOUR PROPOSAL, MR. SALT?

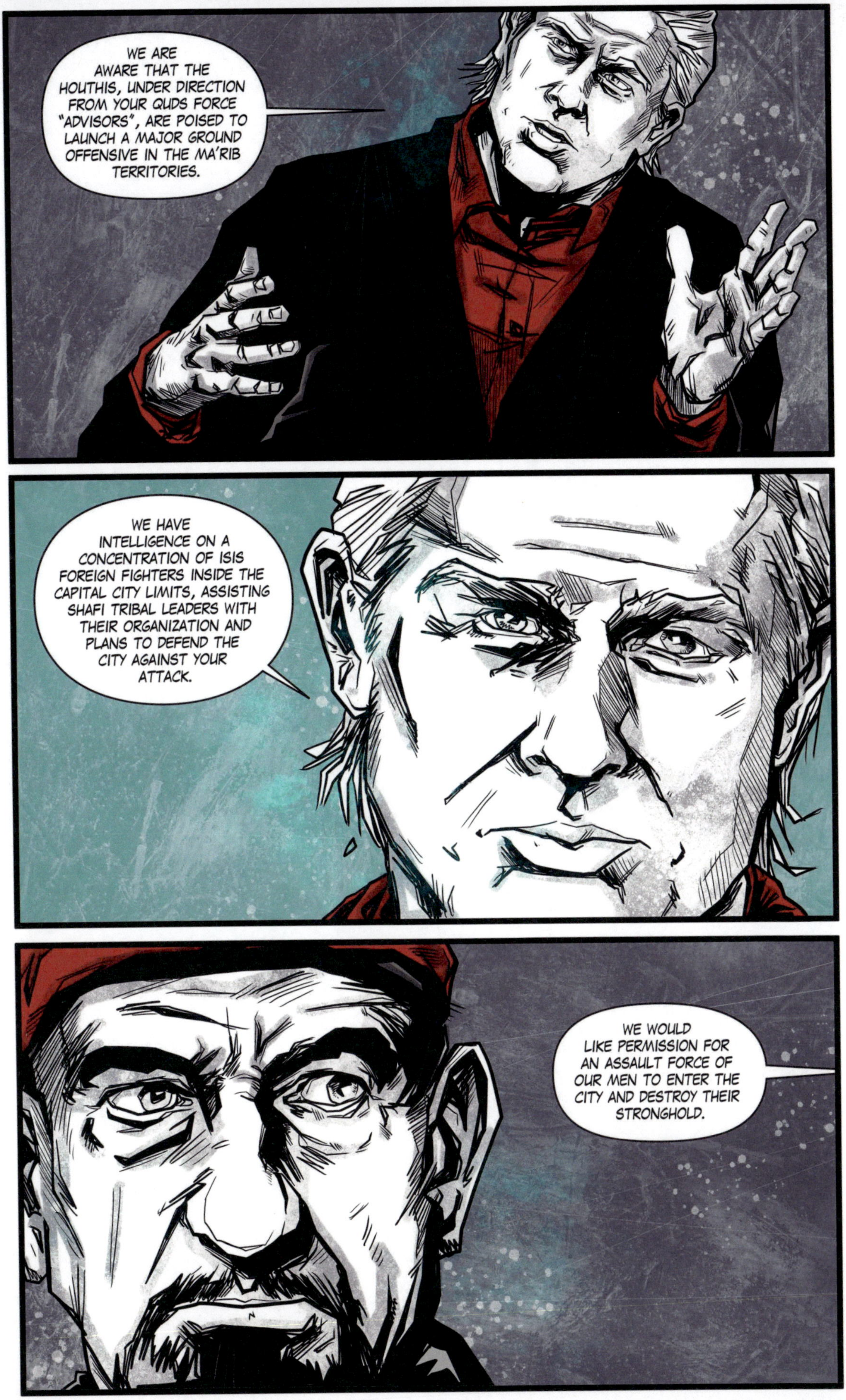
WE ARE AWARE THAT THE HOUTHIS, UNDER DIRECTION FROM YOUR QUDS FORCE "ADVISORS", ARE POISED TO LAUNCH A MAJOR GROUND OFFENSIVE IN THE MA'RIB TERRITORIES.
WE HAVE INTELLIGENCE ON A CONCENTRATION OF ISIS FOREIGN FIGHTERS INSIDE THE CAPITAL CITY LIMITS, ASSISTING SHAFI TRIBAL LEADERS WITH THEIR ORGANIZATION AND PLANS TO DEFEND THE CITY AGAINST YOUR ATTACK.
WE WOULD LIKE PERMISSION FOR AN ASSAULT FORCE OF OUR MEN TO ENTER THE CITY AND DESTROY THEIR STRONGHOLD.

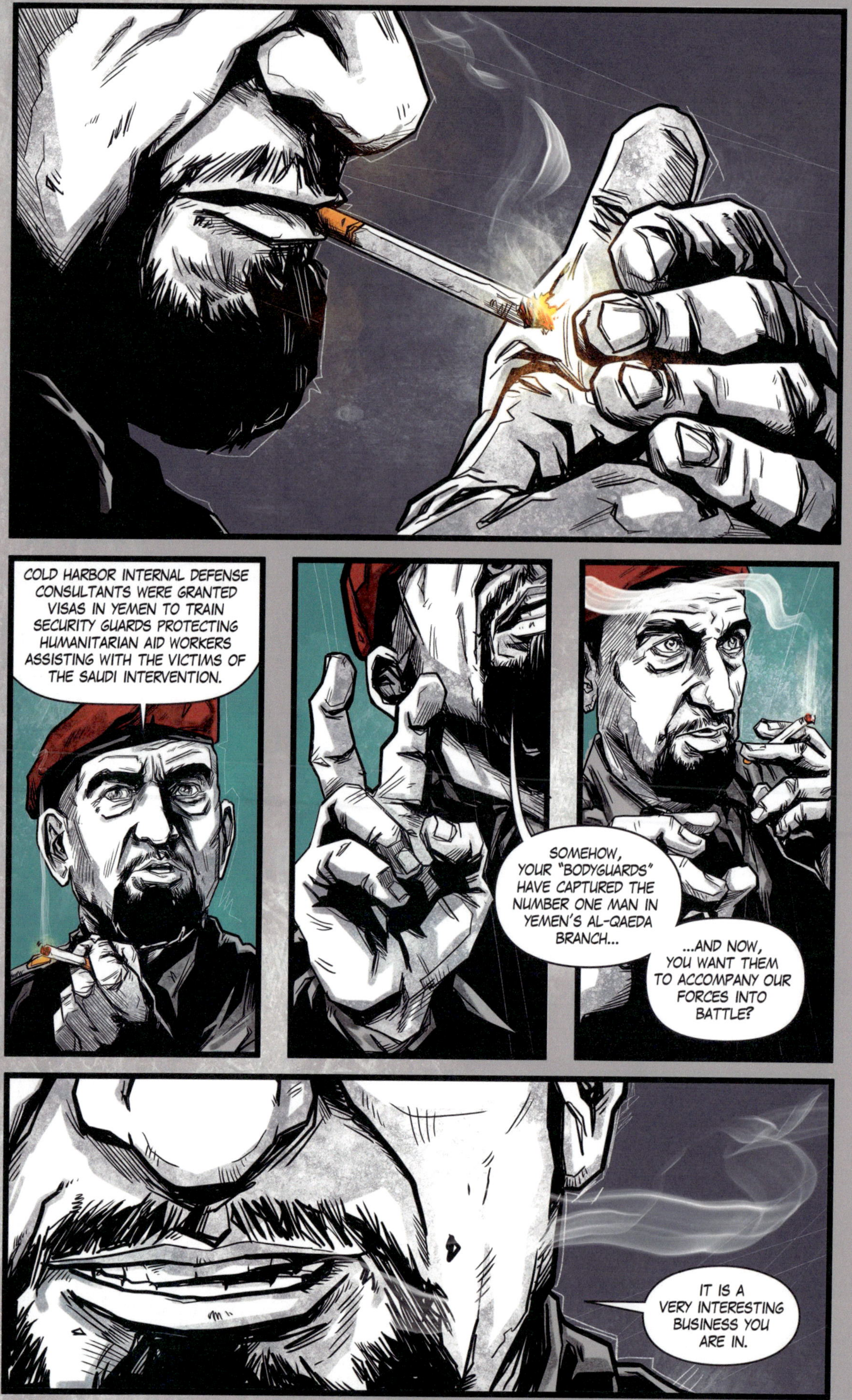
COLD HARBOR INTERNAL DEFENSE CONSULTANTS WERE GRANTED VISAS IN YEMEN TO TRAIN SECURITY GUARDS PROTECTING HUMANITARIAN AID WORKERS ASSISTING WITH THE VICTIMS OF THE SAUDI INTERVENTION.
SOMEHOW, YOUR "BODYGUARDS" HAVE CAPTURED THE NUMBER ONE MAN IN YEMEN'S AL-QAEDA BRANCH...
...AND NOW, YOU WANT THEM TO ACCOMPANY OUR FORCES INTO BATTLE?
IT IS A VERY INTERESTING BUSINESS YOU ARE IN.

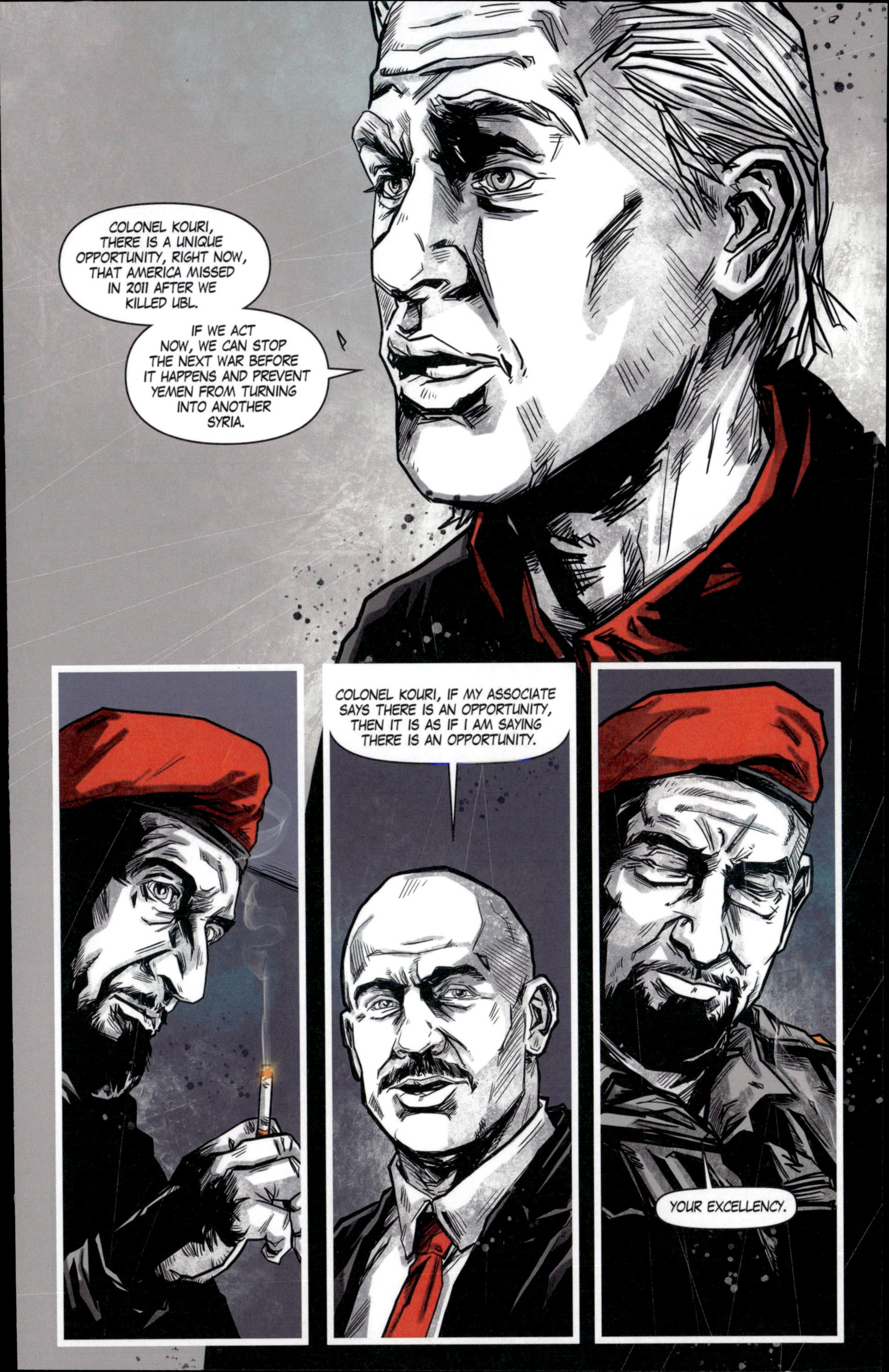
COLONEL KOURI, THERE IS A UNIQUE OPPORTUNITY, RIGHT NOW, THAT AMERICA MISSED IN 2011 AFTER WE KILLED UBL.
IF WE ACT NOW, WE CAN STOP THE NEXT WAR BEFORE IT HAPPENS AND PREVENT YEMEN FROM TURNING INTO ANOTHER SYRIA.
COLONEL KOURI, IF MY ASSOCIATE SAYS THERE IS AN OPPORTUNITY, THEN IT IS AS IF I AM SAYING THERE IS AN OPPORTUNITY.
YOUR EXCELLENCY.

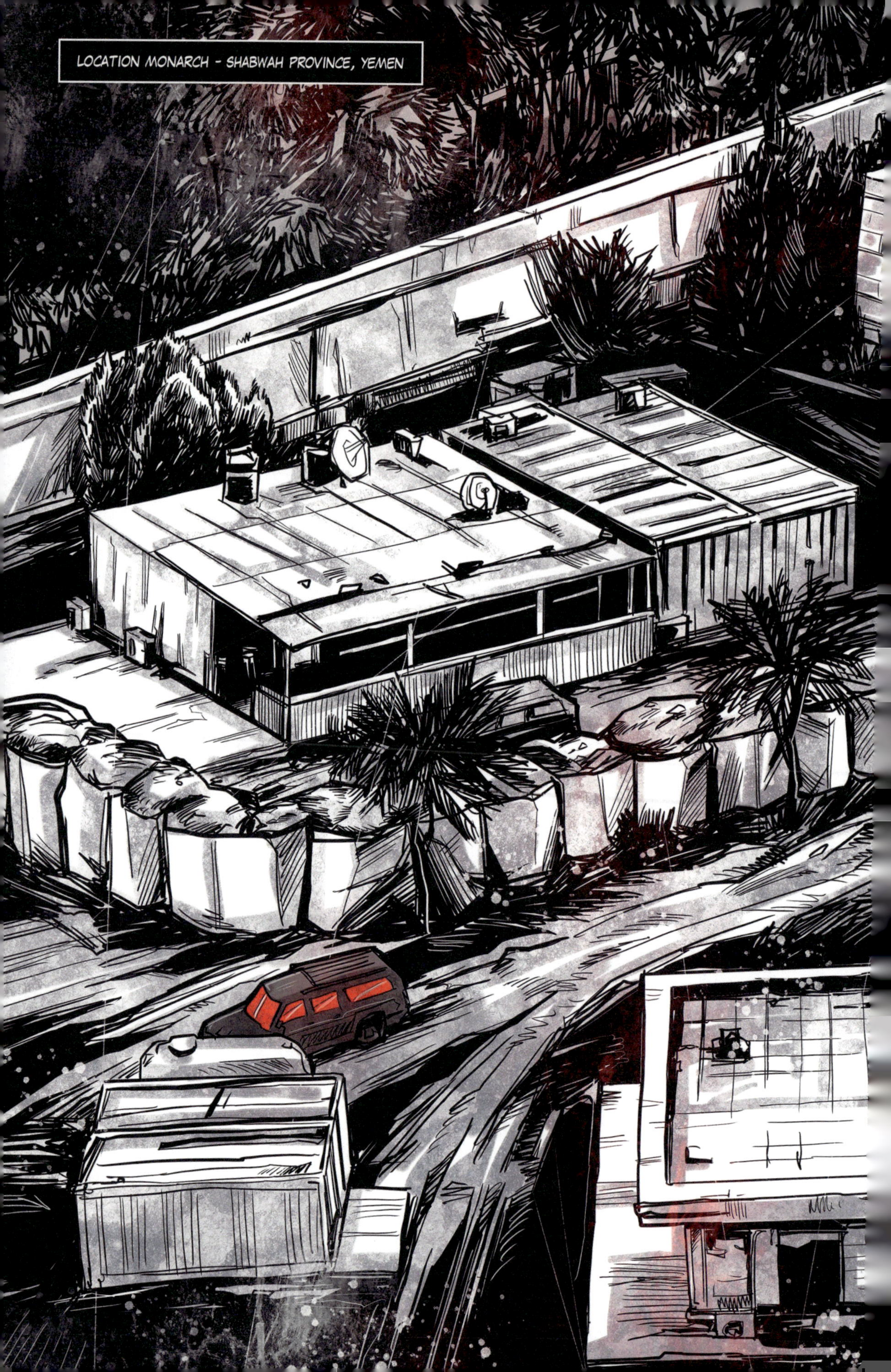
LOCATION MONARCH - SHABWAH PROVINCE, YEMEN

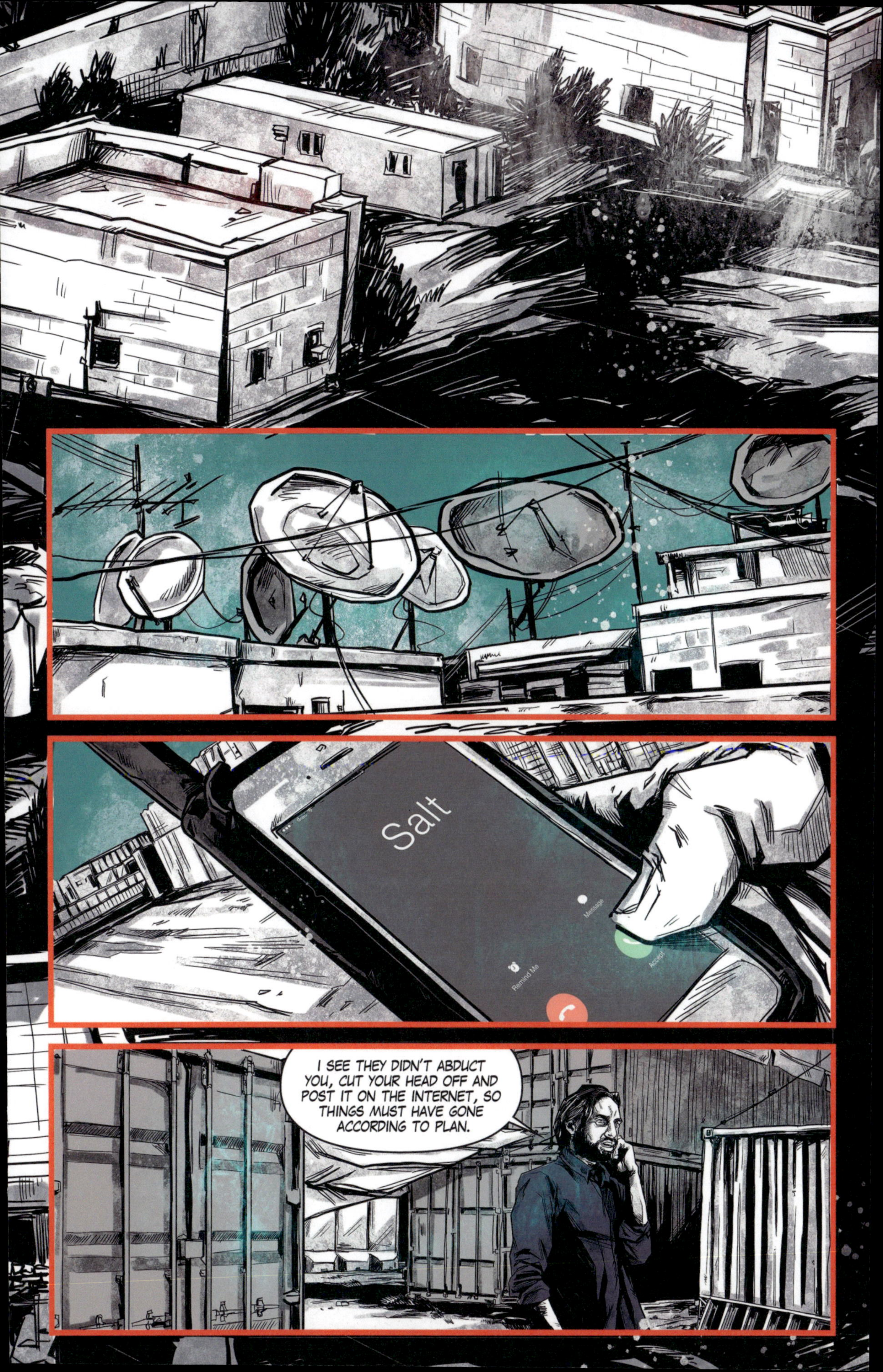
Salt
Message
Remind Me
Accept
I SEE THEY DIDN'T ABDUCT YOU, CUT YOUR HEAD OFF AND POST IT ON THE INTERNET, SO THINGS MUST HAVE GONE ACCORDING TO PLAN.

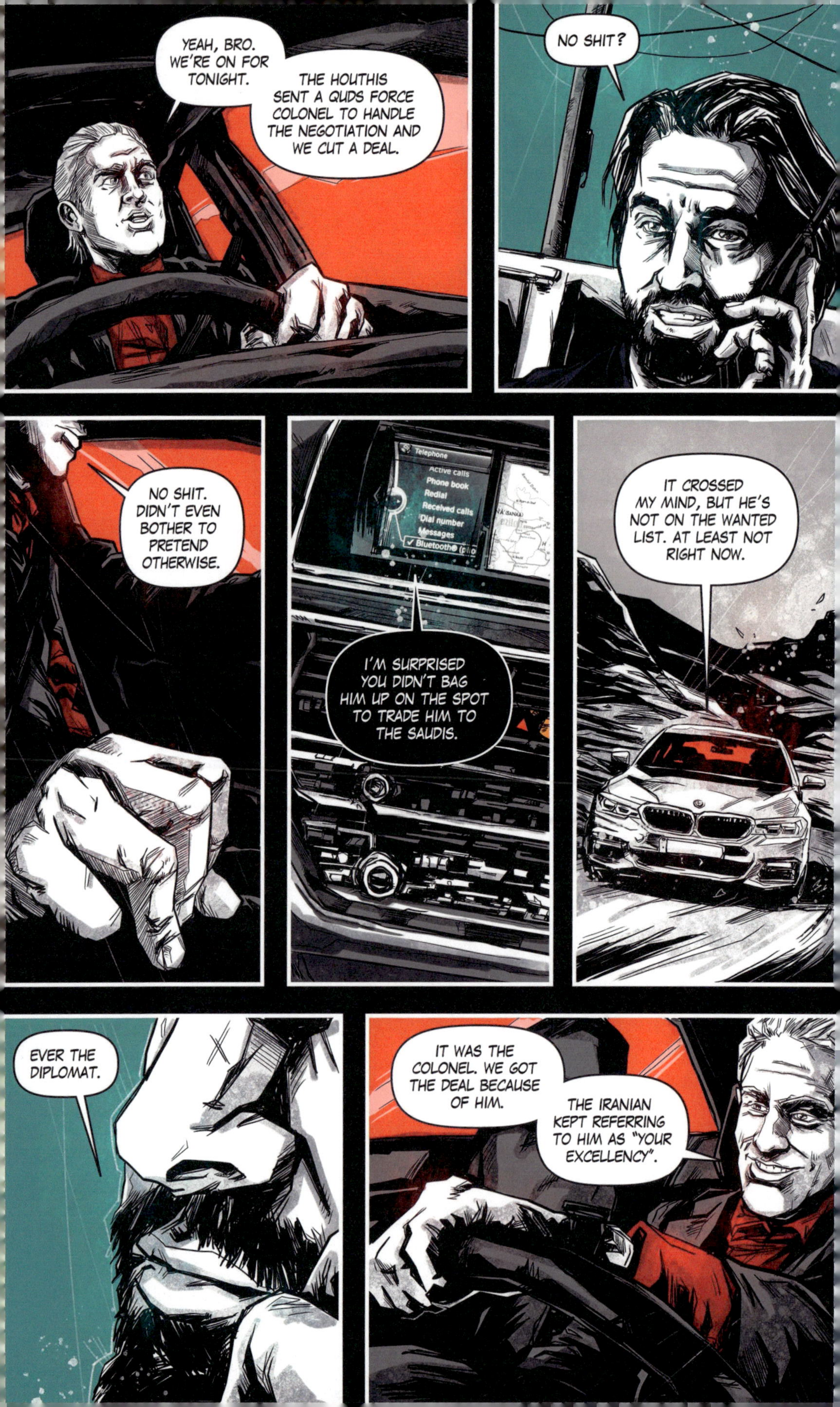
YEAH, BRO. WE'RE ON FOR TONIGHT.
THE HOUTHIS SENT A QUDS FORCE COLONEL TO HANDLE THE NEGOTIATION AND WE CUT A DEAL.
NO SHIT?
NO SHIT. DIDN'T EVEN BOTHER TO PRETEND OTHERWISE.
Telephone
Active calls
Phone book
Redial
Received calls
Dial number
Messages
I'M SURPRISED YOU DIDN'T BAG HIM UP ON THE SPOT TO TRADE HIM TO THE SAUDIS.
IT CROSSED MY MIND, BUT HE'S NOT ON THE WANTED LIST. AT LEAST NOT RIGHT NOW.
EVER THE DIPLOMAT.
IT WAS THE COLONEL. WE GOT THE DEAL BECAUSE OF HIM.
THE IRANIAN KEPT REFERRING TO HIM AS "YOUR EXCELLENCY".

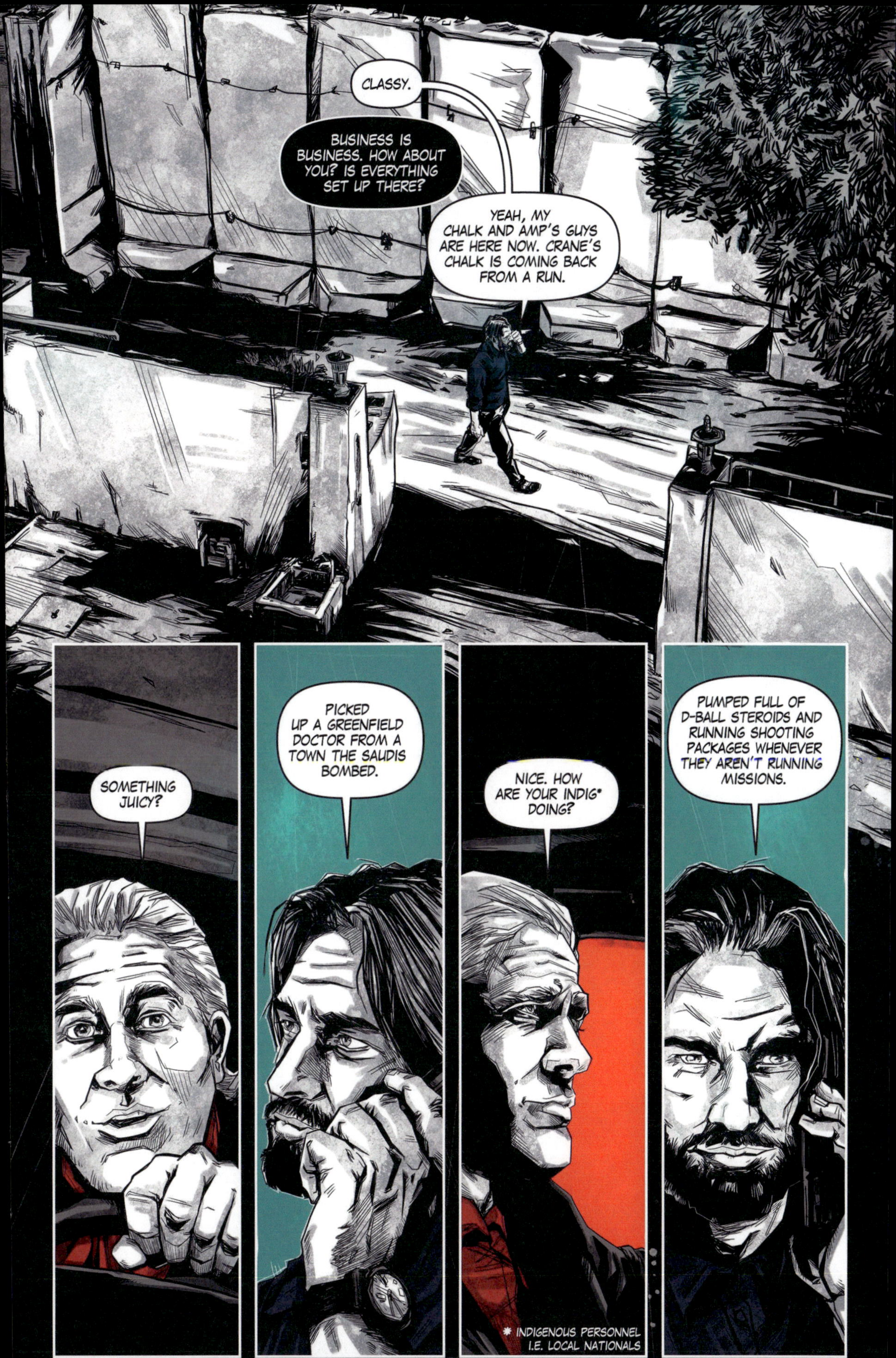
CLASSY.
BUSINESS IS BUSINESS. HOW ABOUT YOU? IS EVERYTHING SET UP THERE?
YEAH, MY CHALK AND AMP'S GUYS ARE HERE NOW. CRANE'S CHALK IS COMING BACK FROM A RUN.
SOMETHING JUICY?
PICKED UP A GREENFIELD DOCTOR FROM A TOWN THE SAUDIS BOMBED.
NICE. HOW ARE YOUR INDIG* DOING?
PUMPED FULL OF D-BALL STEROIDS AND RUNNING SHOOTING PACKAGES WHENEVER THEY AREN'T RUNNING MISSIONS.
* INDIGENOUS PERSONNEL I.E. LOCAL NATIONALS

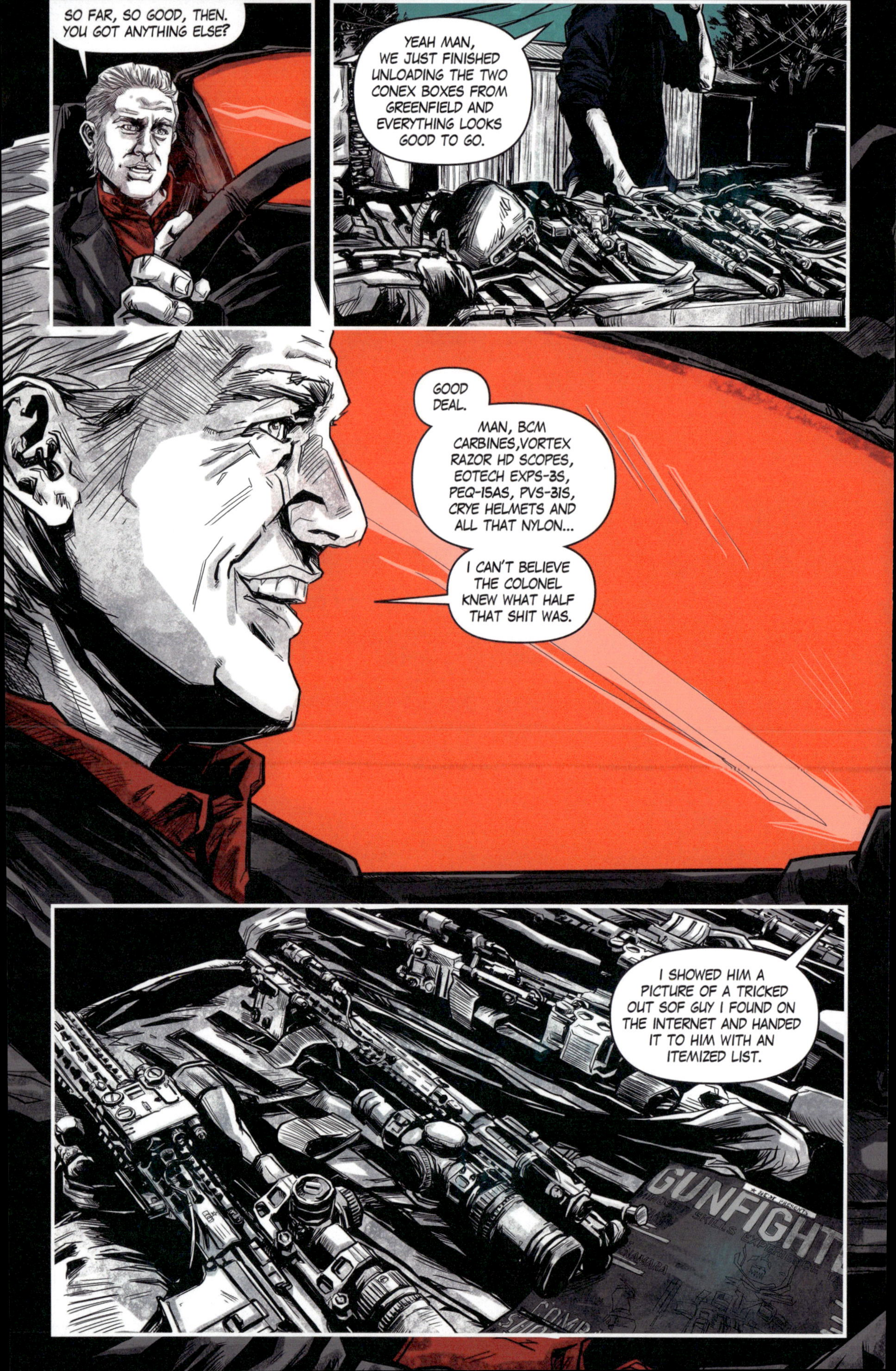
SO FAR, SO GOOD, THEN. YOU GOT ANYTHING ELSE?
YEAH MAN, WE JUST FINISHED UNLOADING THE TWO CONEX BOXES FROM GREENFIELD AND EVERYTHING LOOKS GOOD TO GO.
GOOD DEAL.
MAN, BCM CARBINES,VORTEX RAZOR HD SCOPES, EOTECH EXPS-3S, PEQ-15AS, PVS-31S, CRYE HELMETS AND ALL THAT NYLON...
I CAN'T BELIEVE THE COLONEL KNEW WHAT HALF THAT SHIT WAS.
I SHOWED HIM A PICTURE OF A TRICKED OUT SOF GUY I FOUND ON THE INTERNET AND HANDED IT TO HIM WITH AN ITEMIZED LIST.

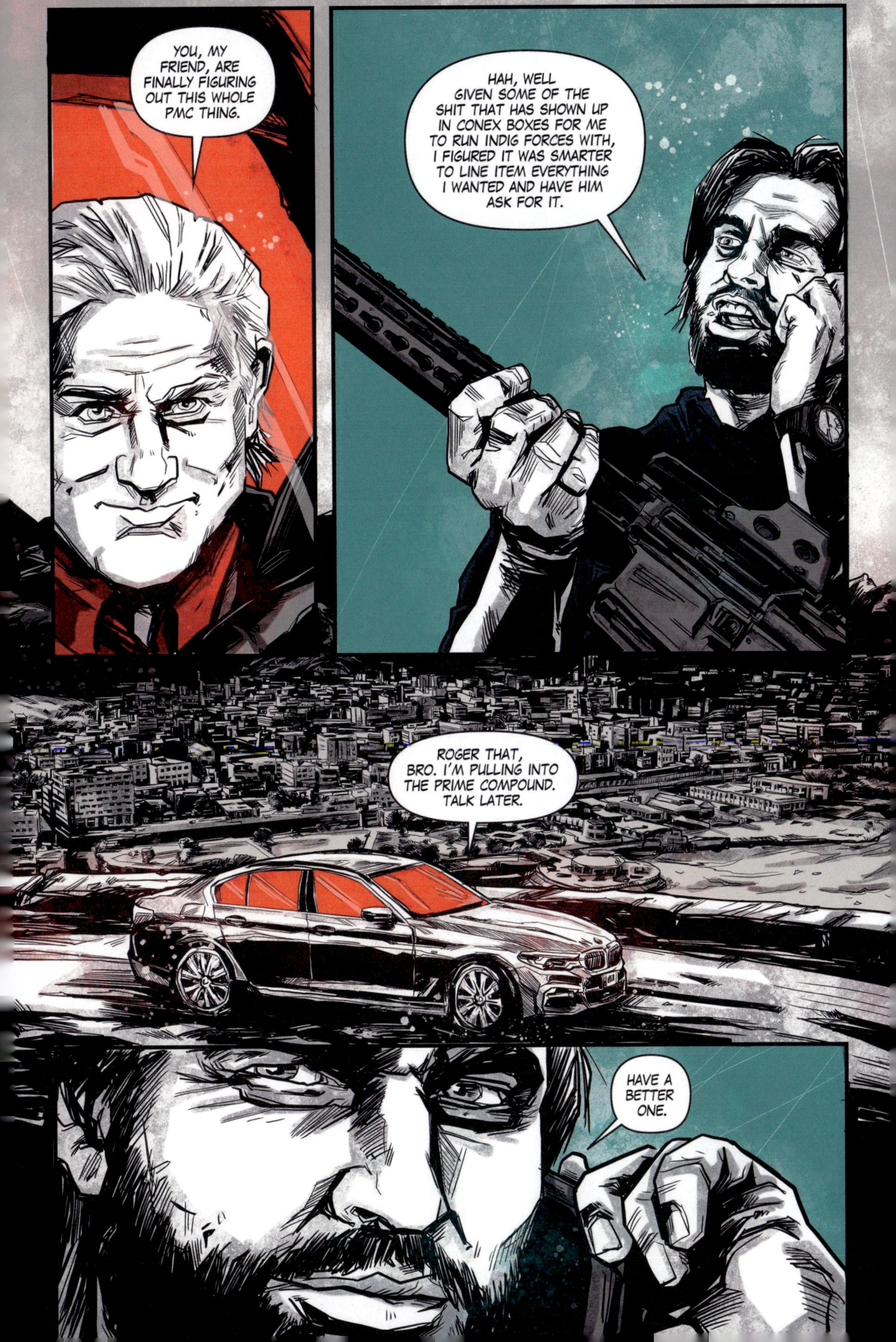
YOU, MY FRIEND, ARE FINALLY FIGURING OUT THIS WHOLE PMC THING.
HAH, WELL GIVEN SOME OF THE SHIT THAT HAS SHOWN UP IN CONEX BOXES FOR ME TO RUN INDIG FORCES WITH, I FIGURED IT WAS SMARTER TO LINE ITEM EVERYTHING I WANTED AND HAVE HIM ASK FOR IT.
ROGER THAT, BRO. I'M PULLING INTO THE PRIME COMPOUND. TALK LATER.
HAVE A BETTER ONE.

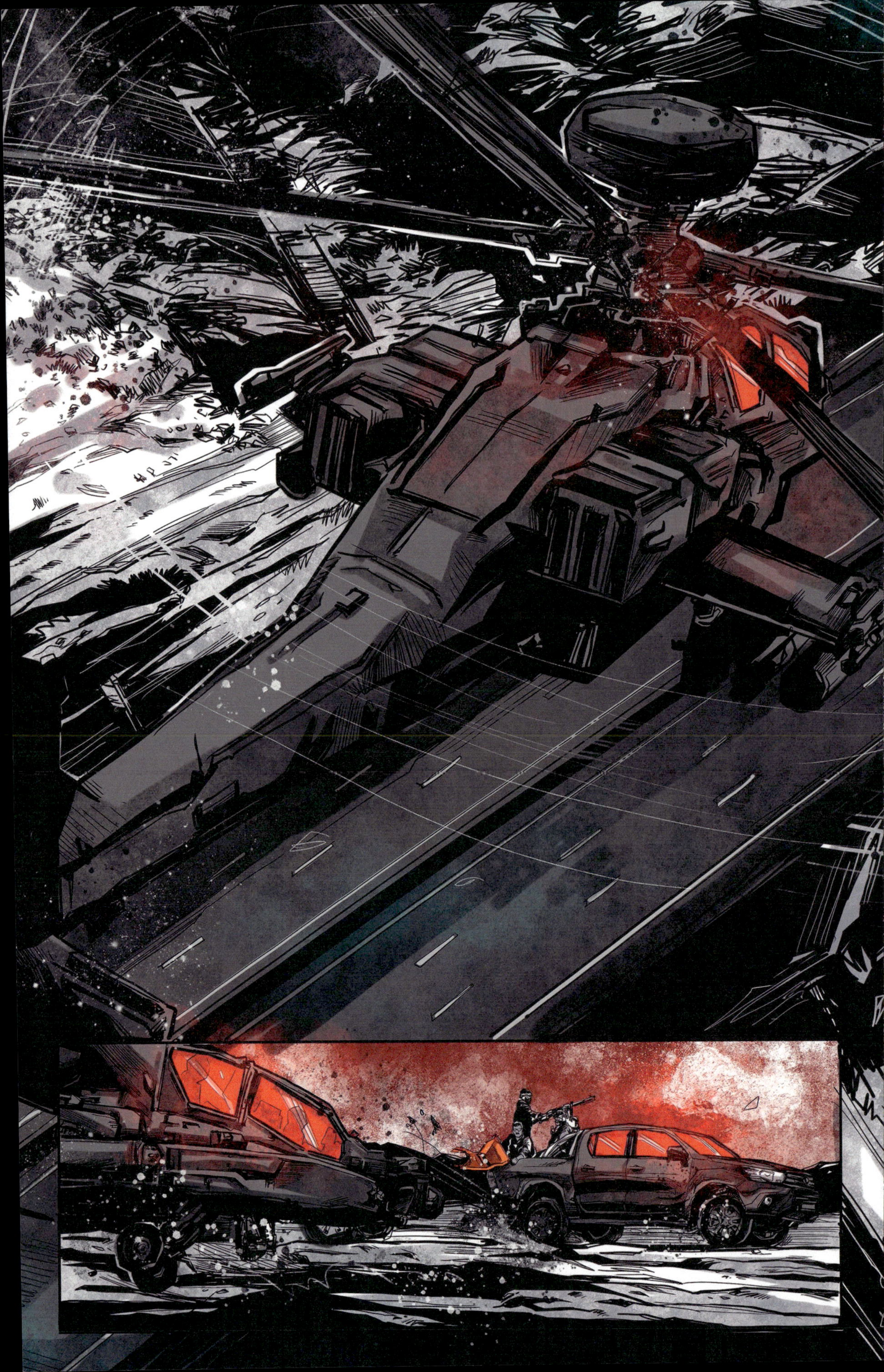

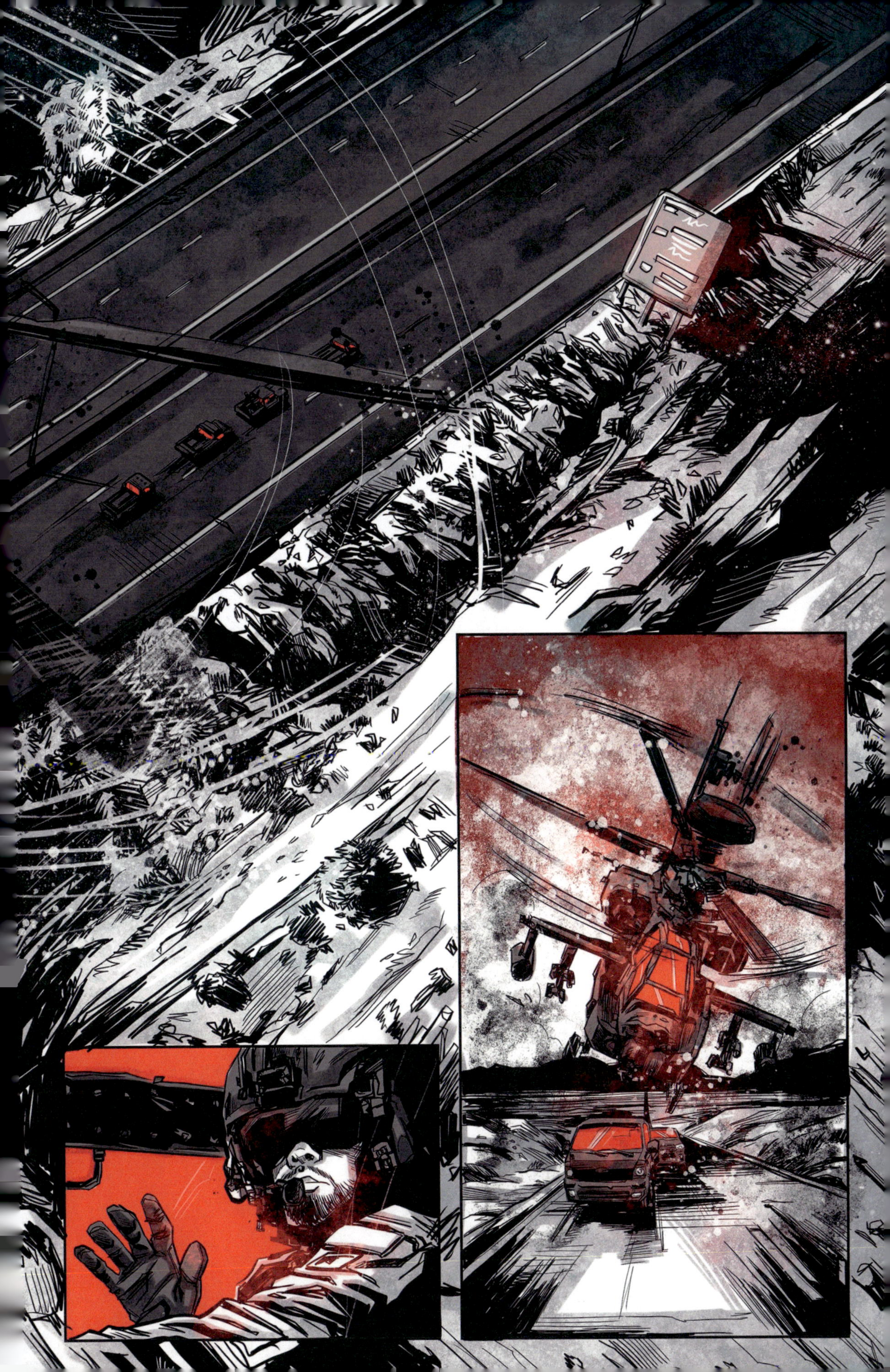

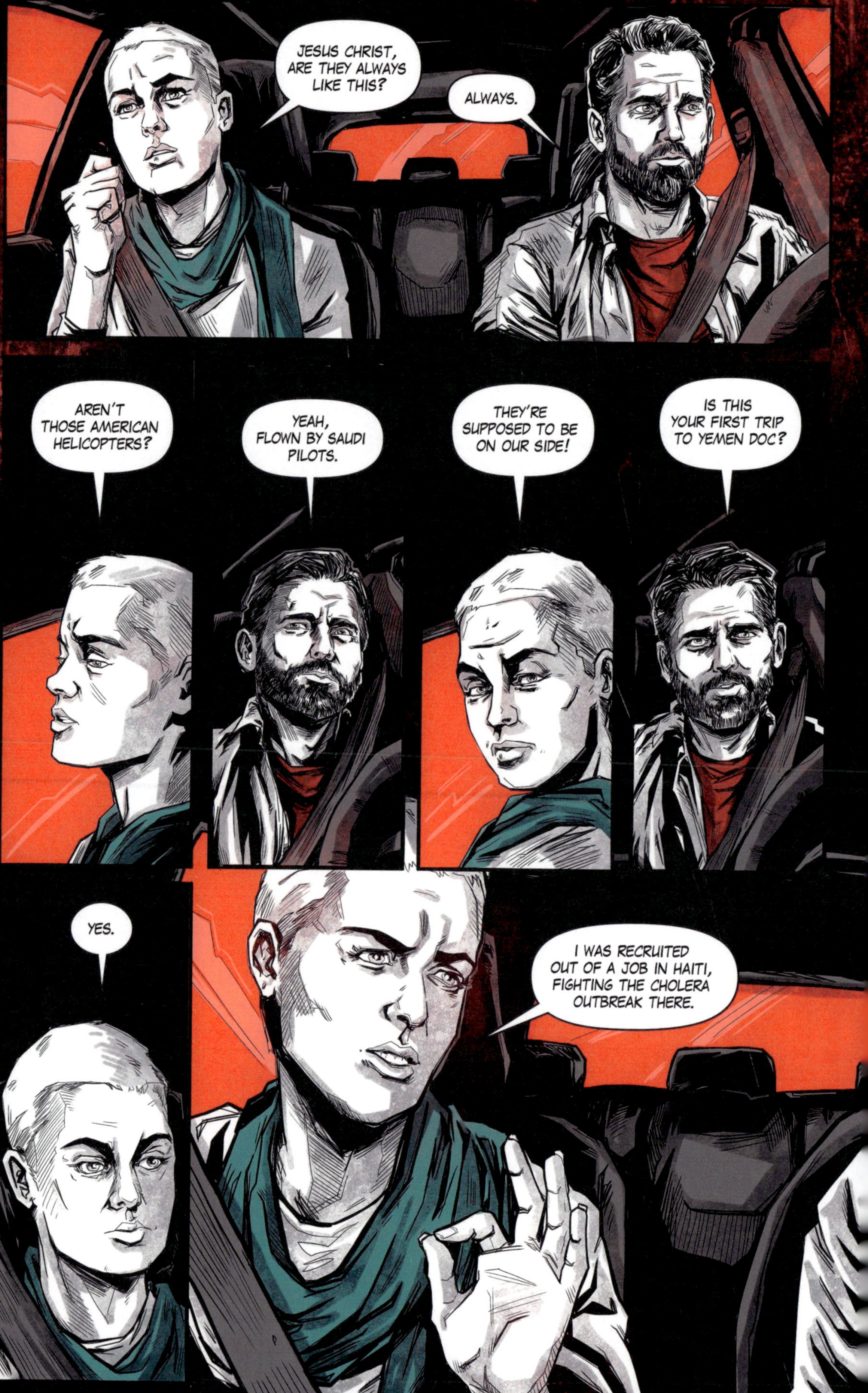
JESUS CHRIST, ARE THEY ALWAYS LIKE THIS?
ALWAYS.
AREN'T THOSE AMERICAN HELICOPTERS?
YEAH, FLOWN BY SAUDI PILOTS.
THEY'RE SUPPOSED TO BE ON OUR SIDE!
IS THIS YOUR FIRST TRIP TO YEMEN DOC?
YES.
I WAS RECRUITED OUT OF A JOB IN HAITI, FIGHTING THE CHOLERA OUTBREAK THERE.

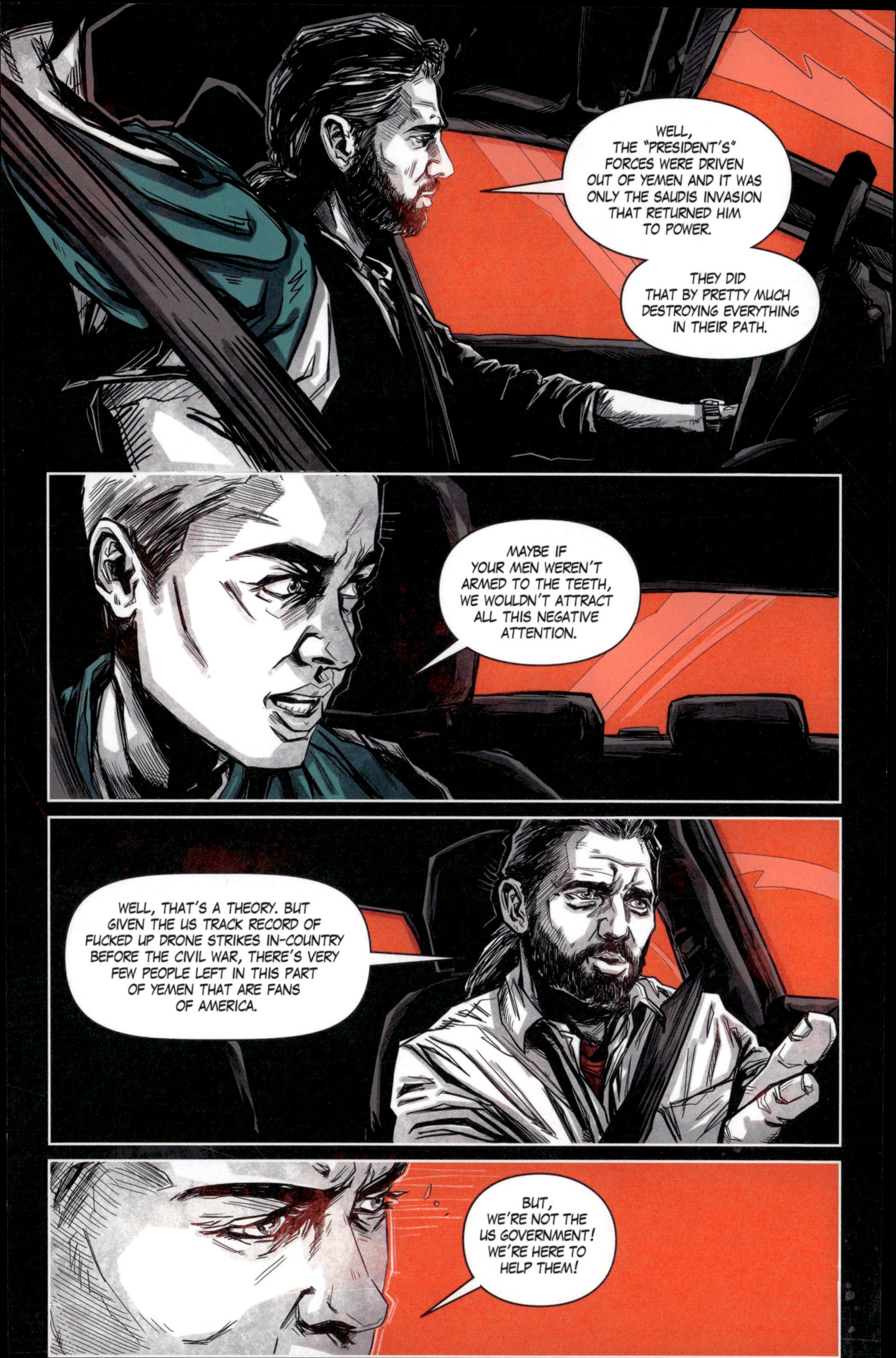
WELL, THE "PRESIDENT'S" FORCES WERE DRIVEN OUT OF YEMEN AND IT WAS ONLY THE SAUDIS INVASION THAT RETURNED HIM TO POWER.
THEY DID THAT BY PRETTY MUCH DESTROYING EVERYTHING IN THEIR PATH.
MAYBE IF YOUR MEN WEREN'T ARMED TO THE TEETH, WE WOULDN'T ATTRACT ALL THIS NEGATIVE ATTENTION.
WELL, THAT'S A THEORY. BUT GIVEN THE US TRACK RECORD OF FUCKED UP DRONE STRIKES IN-COUNTRY BEFORE THE CIVIL WAR, THERE'S VERY FEW PEOPLE LEFT IN THIS PART OF YEMEN THAT ARE FANS OF AMERICA.
BUT, WE'RE NOT THE US GOVERNMENT! WE'RE HERE TO HELP THEM!

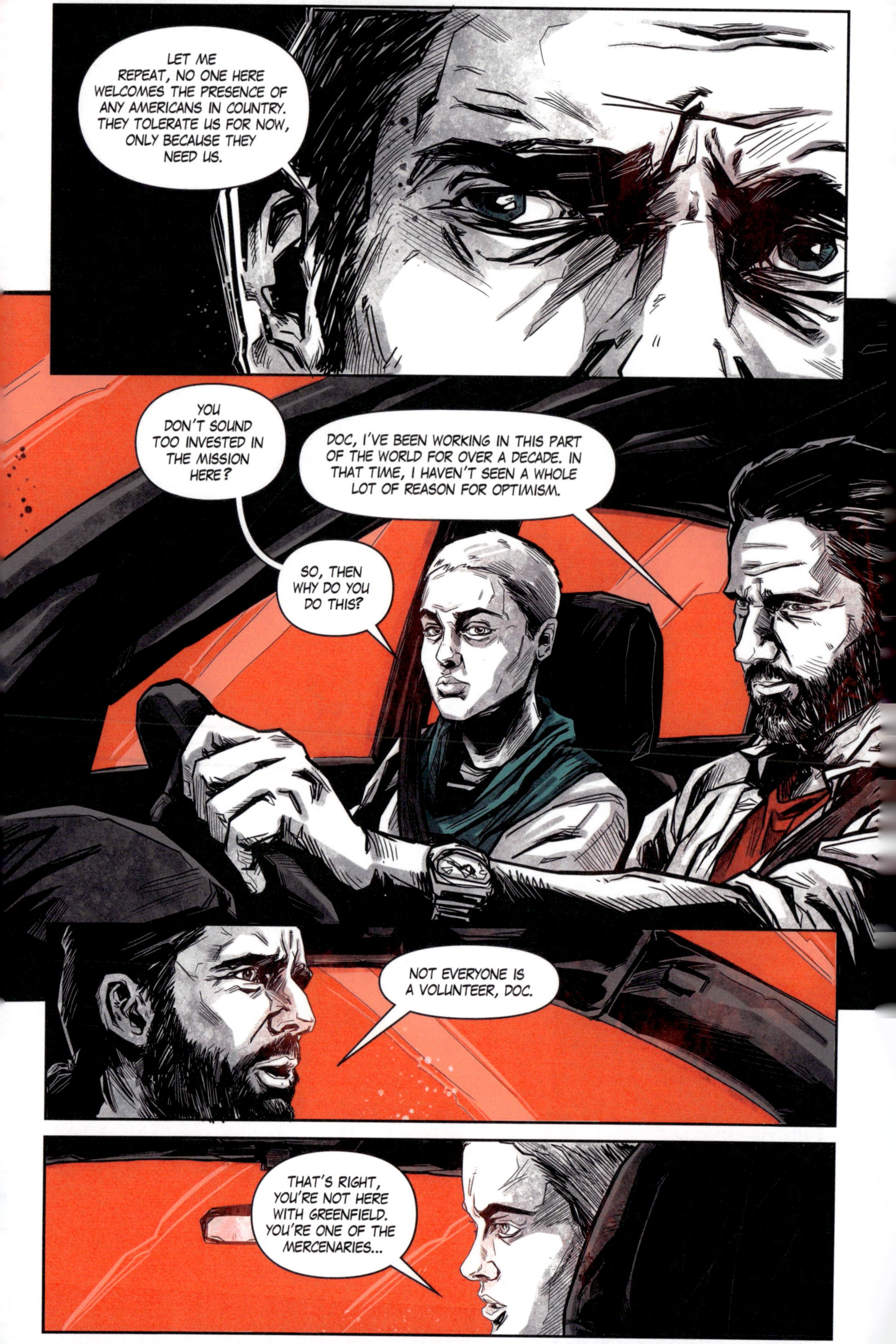
LET ME REPEAT, NO ONE HERE WELCOMES THE PRESENCE OF ANY AMERICANS IN COUNTRY. THEY TOLERATE US FOR NOW, ONLY BECAUSE THEY NEED US.
YOU DON'T SOUND TOO INVESTED IN THE MISSION HERE?
DOC, I'VE BEEN WORKING IN THIS PART OF THE WORLD FOR OVER A DECADE. IN THAT TIME, I HAVEN'T SEEN A WHOLE LOT OF REASON FOR OPTIMISM.
SO, THEN WHY DO YOU DO THIS?
NOT EVERYONE IS A VOLUNTEER, DOC.
THAT'S RIGHT, YOU'RE NOT HERE WITH GREENFIELD. YOU'RE ONE OF THE MERCENARIES...

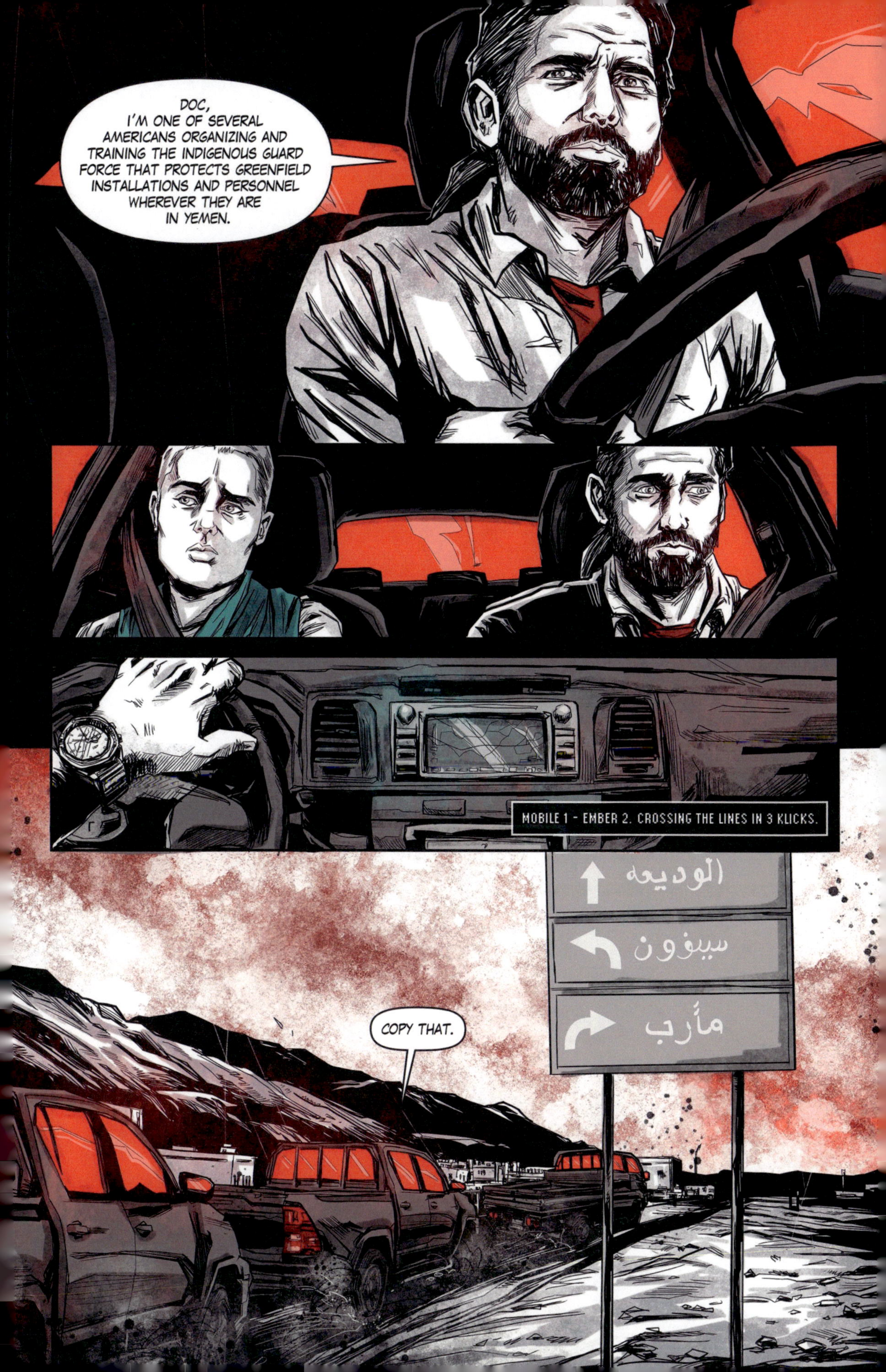
DOC, I'M ONE OF SEVERAL AMERICANS ORGANIZING AND TRAINING THE INDIGENOUS GUARD FORCE THAT PROTECTS GREENFIELD INSTALLATIONS AND PERSONNEL WHEREVER THEY ARE IN YEMEN.
MOBILE 1 - EMBER 2. CROSSING THE LINES IN 3 KLICKS.
الوديعه
سيئون
مأرب
COPY THAT.

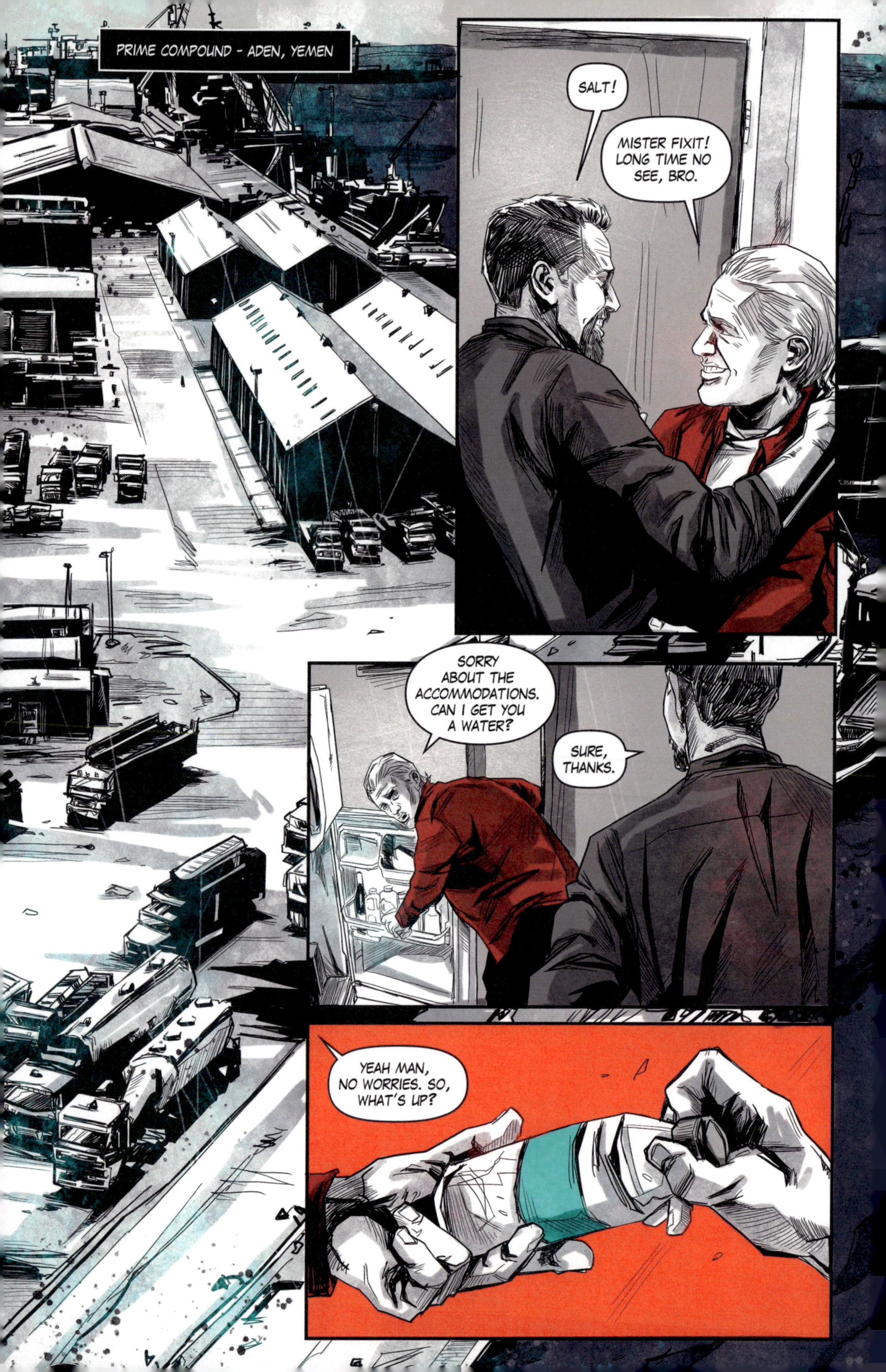
PRIME COMPOUND - ADEN, YEMEN
SALT!
MISTER FIXIT! LONG TIME NO SEE, BRO.
SORRY ABOUT THE ACCOMMODATIONS. CAN I GET YOU A WATER?
SURE, THANKS.
YEAH MAN, NO WORRIES. SO, WHAT'S UP?

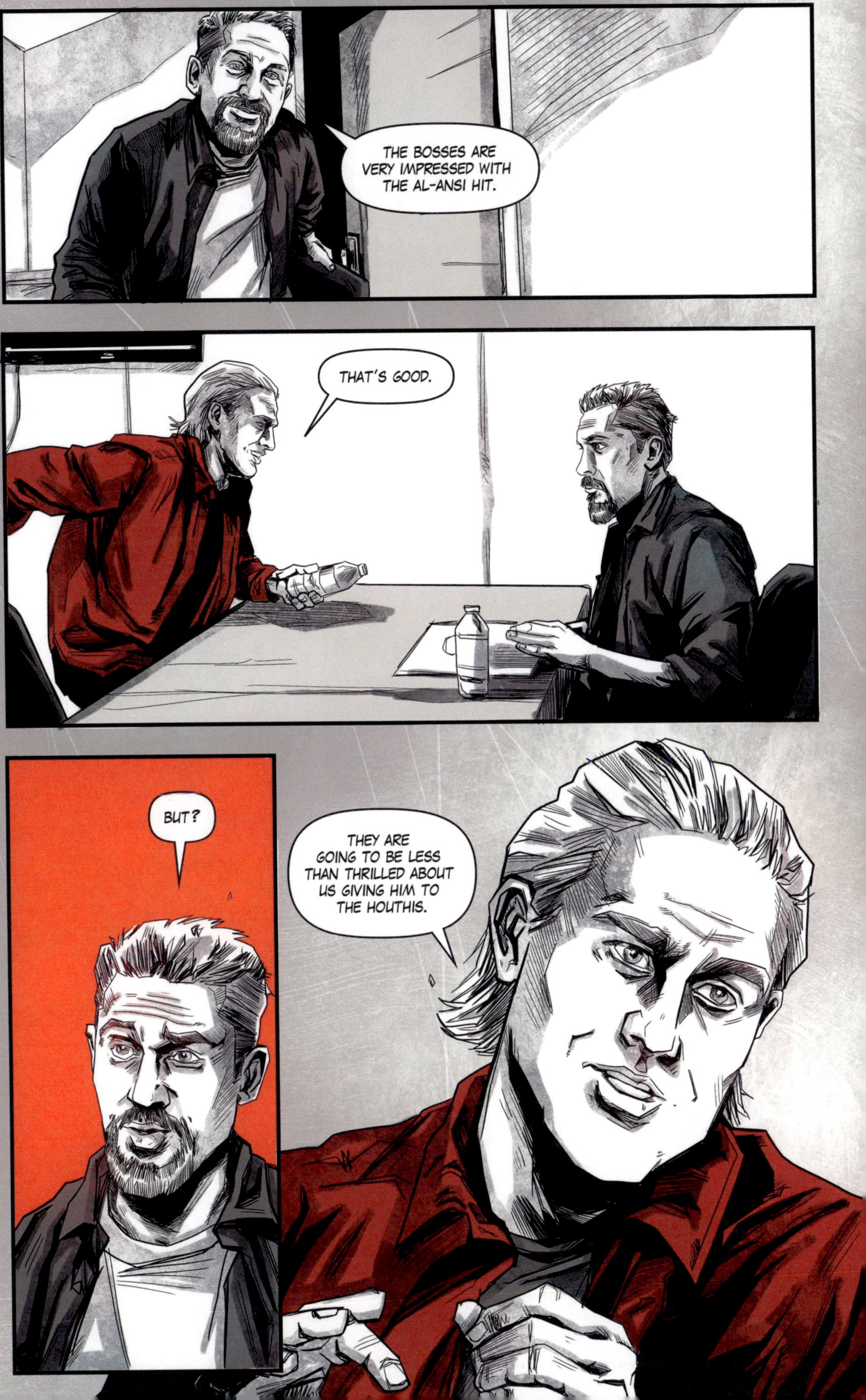
THE BOSSES ARE VERY IMPRESSED WITH THE AL-ANSI HIT.
THAT'S GOOD.
BUT?
THEY ARE GOING TO BE LESS THAN THRILLED ABOUT US GIVING HIM TO THE HOUTHIS.

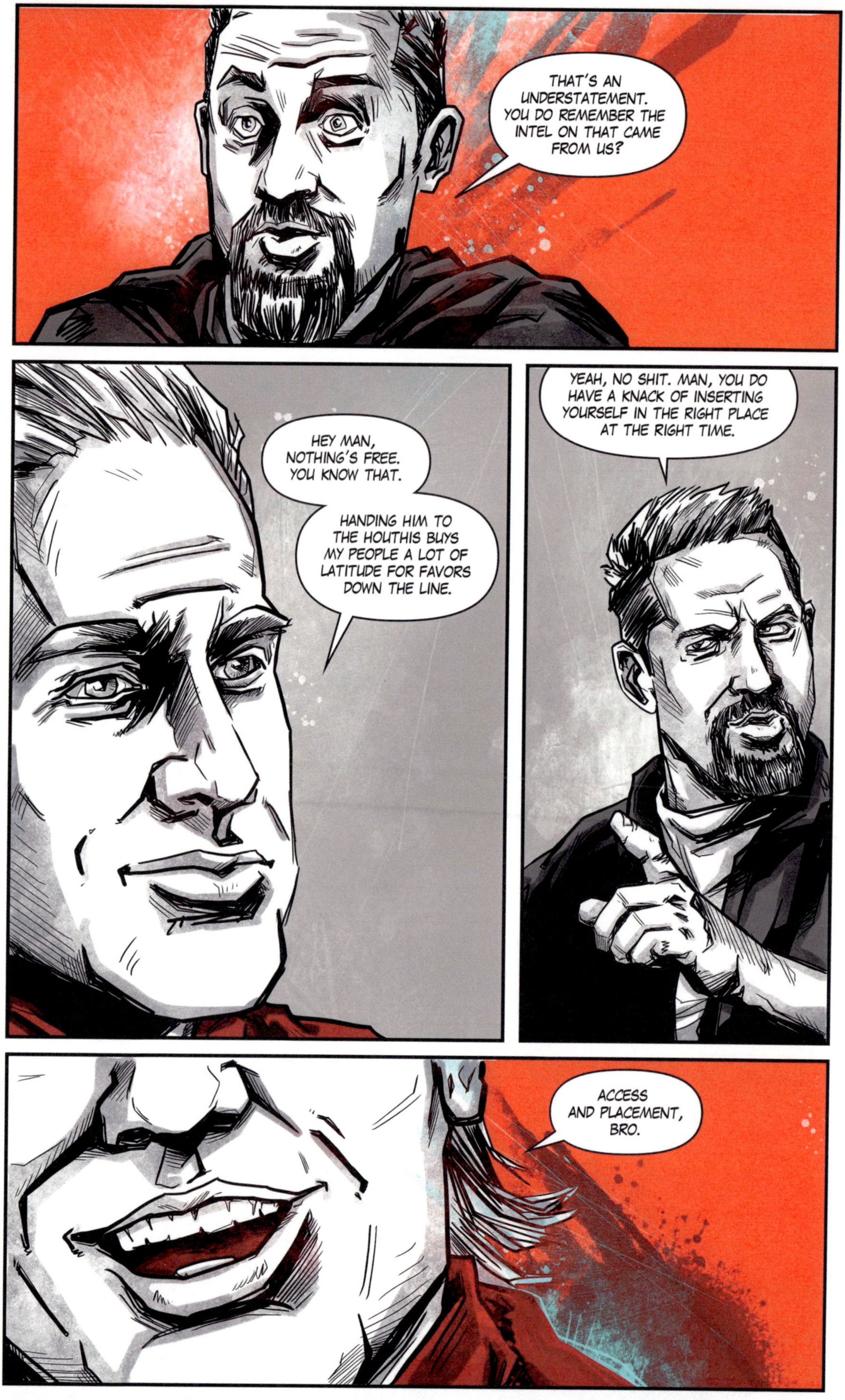
THAT'S AN UNDERSTATEMENT. YOU DO REMEMBER THE INTEL ON THAT CAME FROM US?
HEY MAN, NOTHING'S FREE. YOU KNOW THAT.
HANDING HIM TO THE HOUTHIS BUYS MY PEOPLE A LOT OF LATITUDE FOR FAVORS DOWN THE LINE.
YEAH, NO SHIT. MAN, YOU DO HAVE A KNACK OF INSERTING YOURSELF IN THE RIGHT PLACE AT THE RIGHT TIME.
ACCESS AND PLACEMENT, BRO.

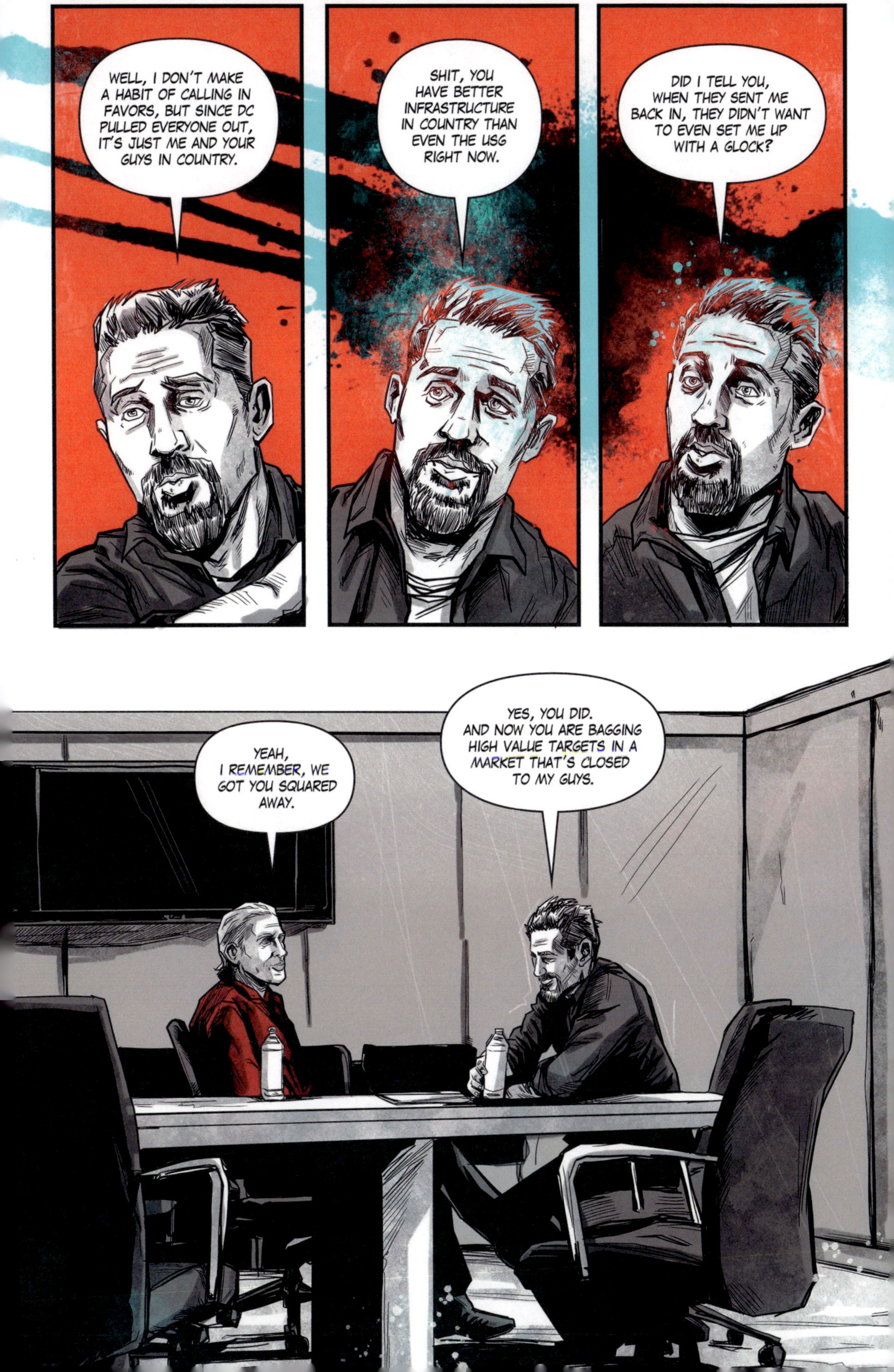
WELL, I DON'T MAKE A HABIT OF CALLING IN FAVORS, BUT SINCE DC PULLED EVERYONE OUT, IT'S JUST ME AND YOUR GUYS IN COUNTRY.
SHIT, YOU HAVE BETTER INFRASTRUCTURE IN COUNTRY THAN EVEN THE USG RIGHT NOW.
DID I TELL YOU, WHEN THEY SENT ME BACK IN, THEY DIDN'T WANT TO EVEN SET ME UP WITH A GLOCK?
YEAH, I REMEMBER, WE GOT YOU SQUARED AWAY.
YES, YOU DID. AND NOW YOU ARE BAGGING HIGH VALUE TARGETS IN A MARKET THAT'S CLOSED TO MY GUYS.

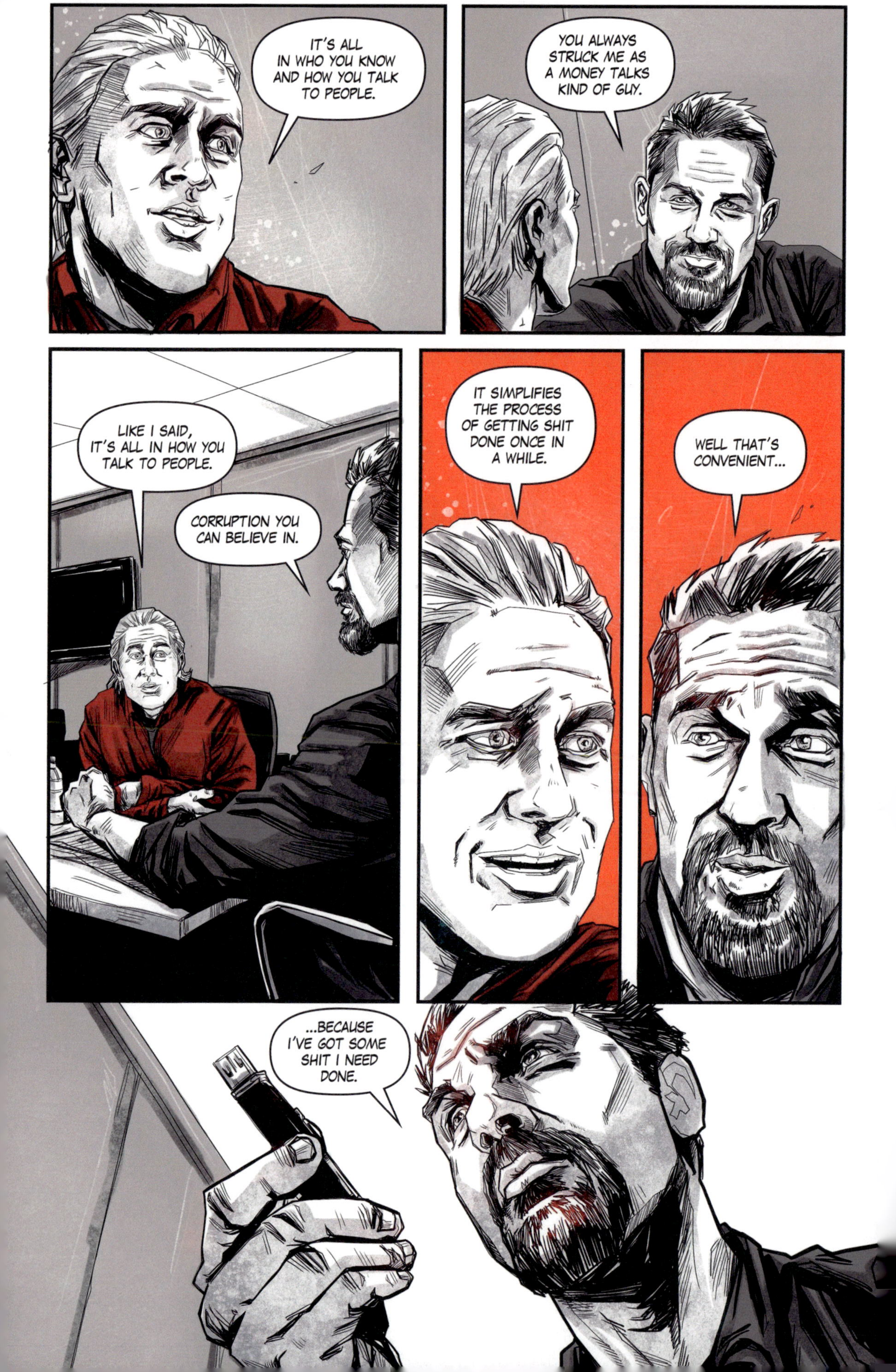
IT'S ALL IN WHO YOU KNOW AND HOW YOU TALK TO PEOPLE.
YOU ALWAYS STRUCK ME AS A MONEY TALKS KIND OF GUY.
LIKE I SAID, IT'S ALL IN HOW YOU TALK TO PEOPLE.
CORRUPTION YOU CAN BELIEVE IN.
IT SIMPLIFIES THE PROCESS OF GETTING SHIT DONE ONCE IN A WHILE.
WELL THAT'S CONVENIENT...
...BECAUSE I'VE GOT SOME SHIT I NEED DONE.

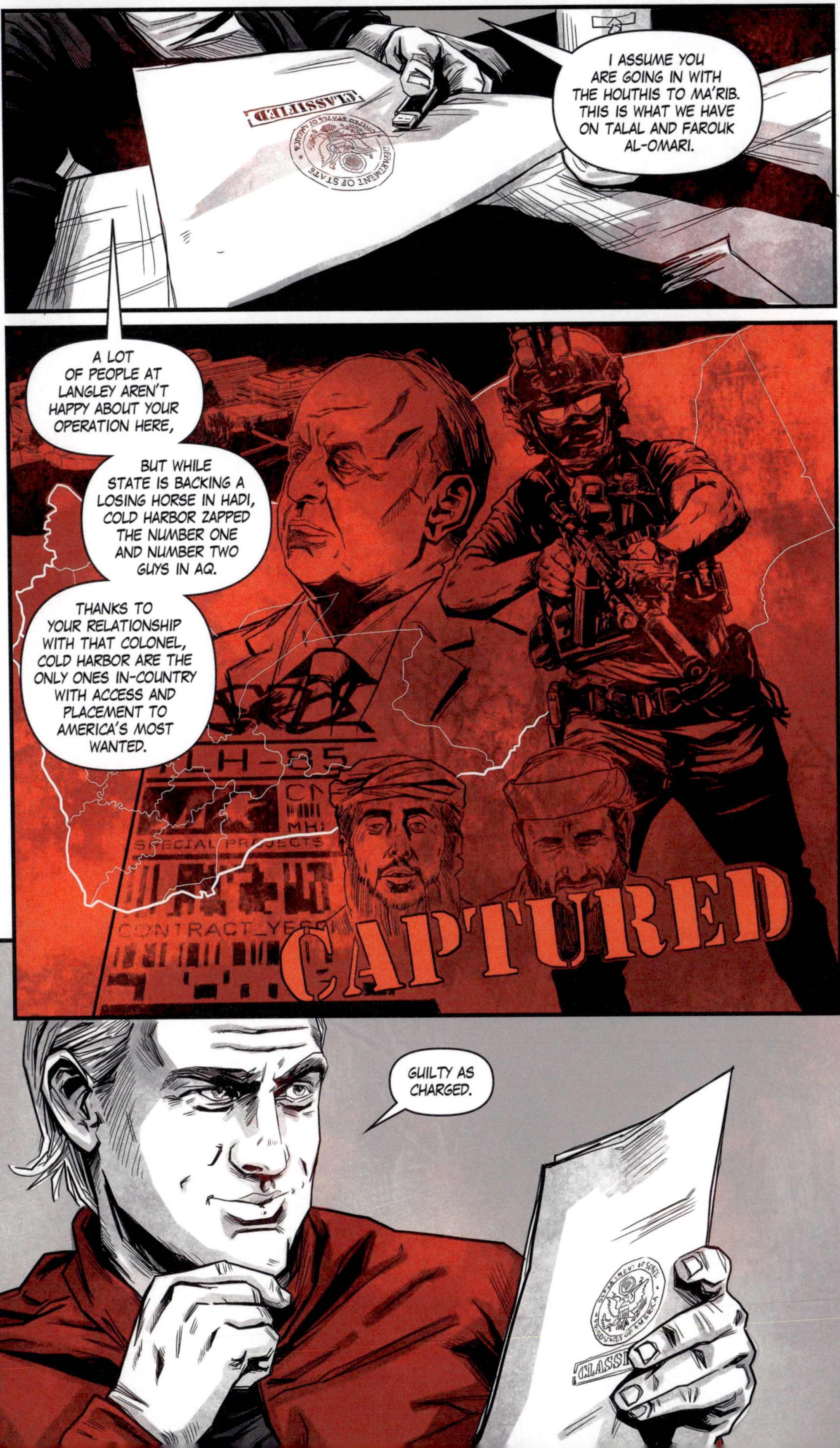

I ASSUME YOU ARE GOING IN WITH THE HOUTHIS TO MA'RIB. THIS IS WHAT WE HAVE ON TALAL AND FAROUK AL-OMARI.
A LOT OF PEOPLE AT LANGLEY AREN'T HAPPY ABOUT YOUR OPERATION HERE,
BUT WHILE STATE IS BACKING A LOSING HORSE IN HADI, COLD HARBOR ZAPPED THE NUMBER ONE AND NUMBER TWO GUYS IN AQ.
THANKS TO YOUR RELATIONSHIP WITH THAT COLONEL, COLD HARBOR ARE THE ONLY ONES IN-COUNTRY WITH ACCESS AND PLACEMENT TO AMERICA'S MOST WANTED.
CAPTURED
GUILTY AS CHARGED.

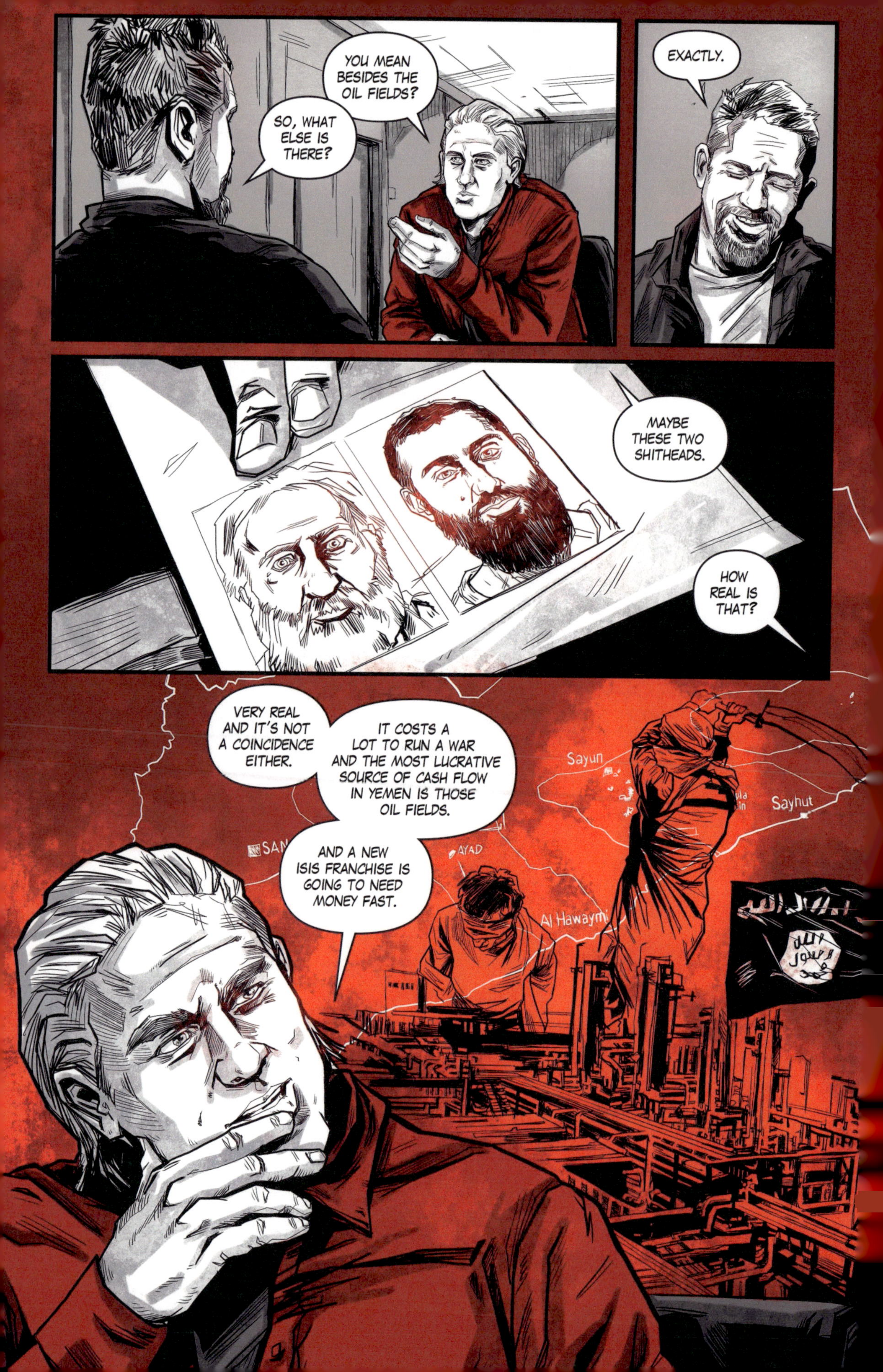

SO, WHAT ELSE IS THERE?
YOU MEAN BESIDES THE OIL FIELDS?
EXACTLY.
MAYBE THESE TWO SHITHEADS.
HOW REAL IS THAT?
VERY REAL AND IT'S NOT A COINCIDENCE EITHER.
IT COSTS A LOT TO RUN A WAR AND THE MOST LUCRATIVE SOURCE OF CASH FLOW IN YEMEN IS THOSE OIL FIELDS.
AND A NEW ISIS FRANCHISE IS GOING TO NEED MONEY FAST.
Sayun
Sayhut
AYAD
Al Hawaymi

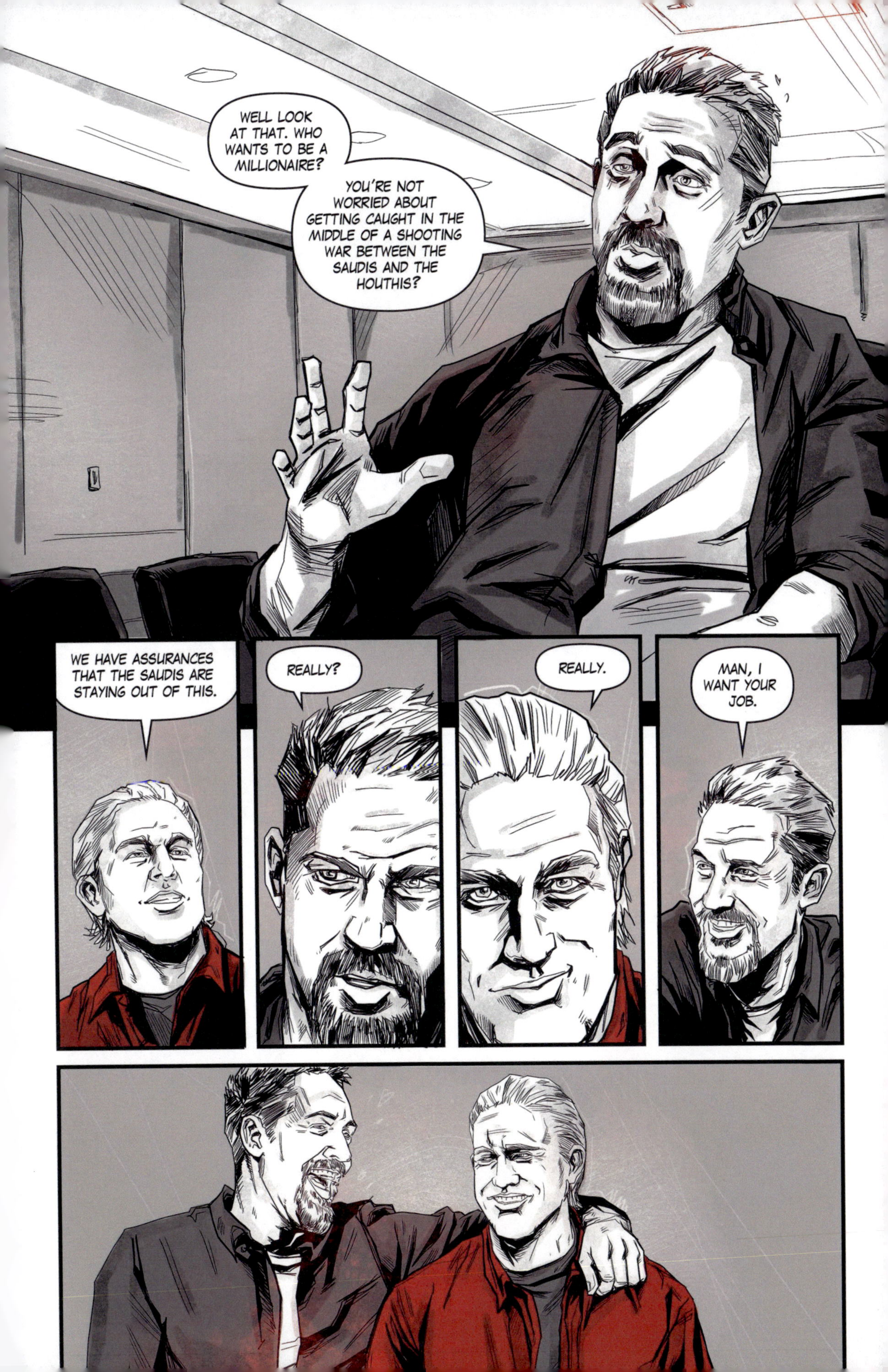
WELL LOOK AT THAT. WHO WANTS TO BE A MILLIONAIRE?
YOU'RE NOT WORRIED ABOUT GETTING CAUGHT IN THE MIDDLE OF A SHOOTING WAR BETWEEN THE SAUDIS AND THE HOUTHIS?
WE HAVE ASSURANCES THAT THE SAUDIS ARE STAYING OUT OF THIS.
REALLY?
REALLY.
MAN, I WANT YOUR JOB.

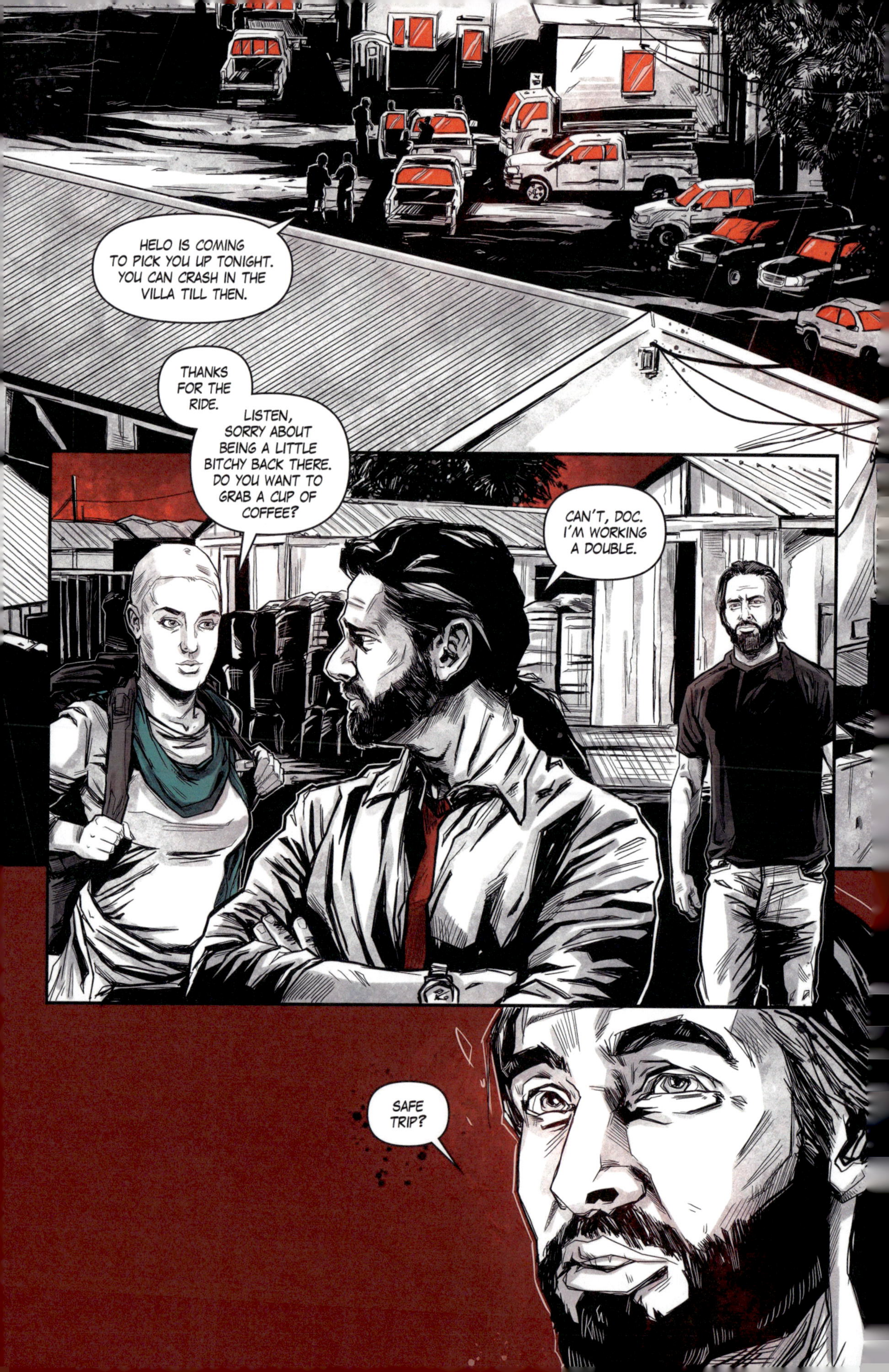
HELO IS COMING TO PICK YOU UP TONIGHT. YOU CAN CRASH IN THE VILLA TILL THEN.
THANKS FOR THE RIDE.
LISTEN, SORRY ABOUT BEING A LITTLE BITCHY BACK THERE. DO YOU WANT TO GRAB A CUP OF COFFEE?
CAN'T, DOC. I'M WORKING A DOUBLE.
SAFE TRIP?

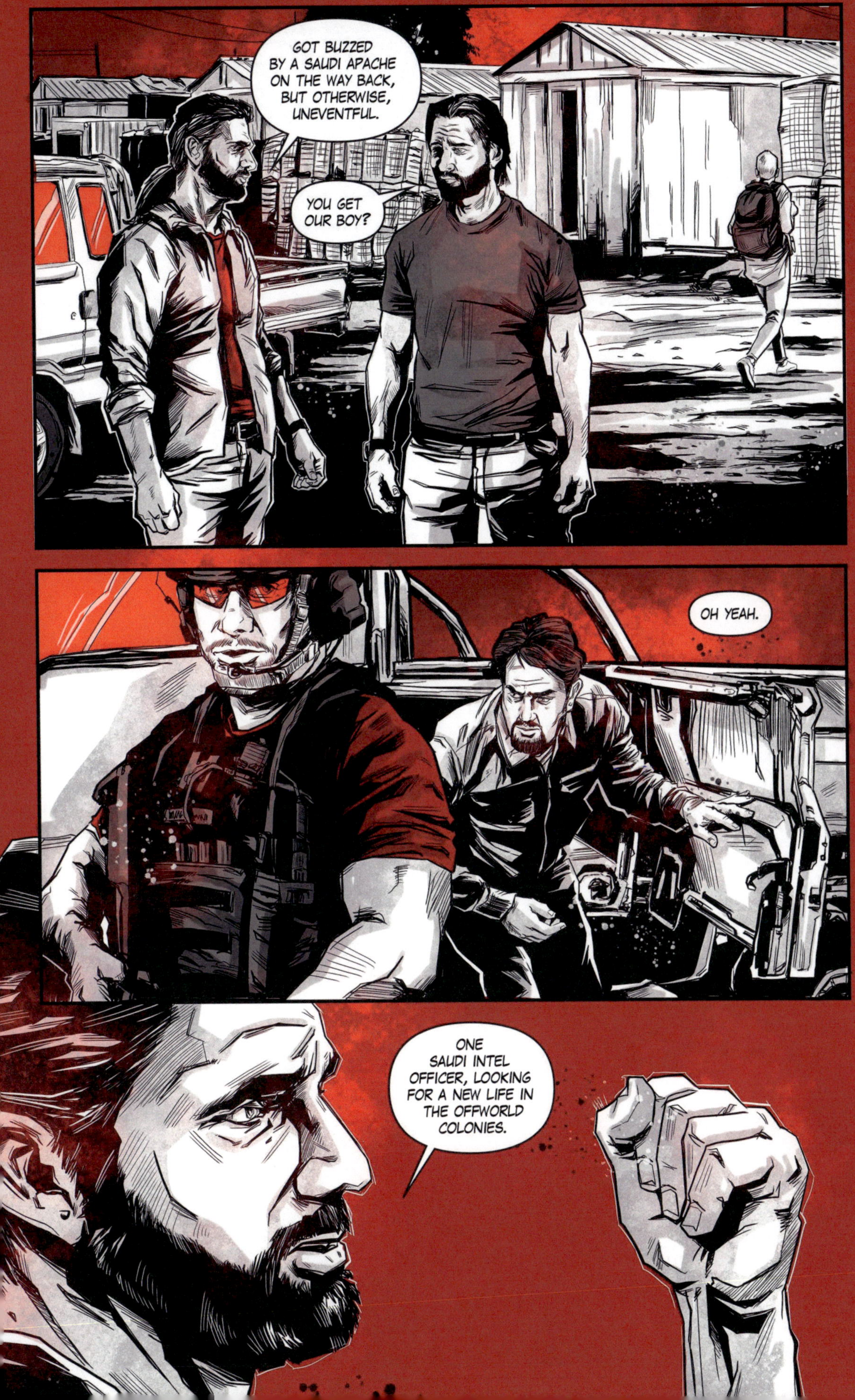
GOT BUZZED BY A SAUDI APACHE ON THE WAY BACK, BUT OTHERWISE, UNEVENTFUL.
YOU GET OUR BOY?
OH YEAH.
ONE SAUDI INTEL OFFICER, LOOKING FOR A NEW LIFE IN THE OFFWORLD COLONIES.

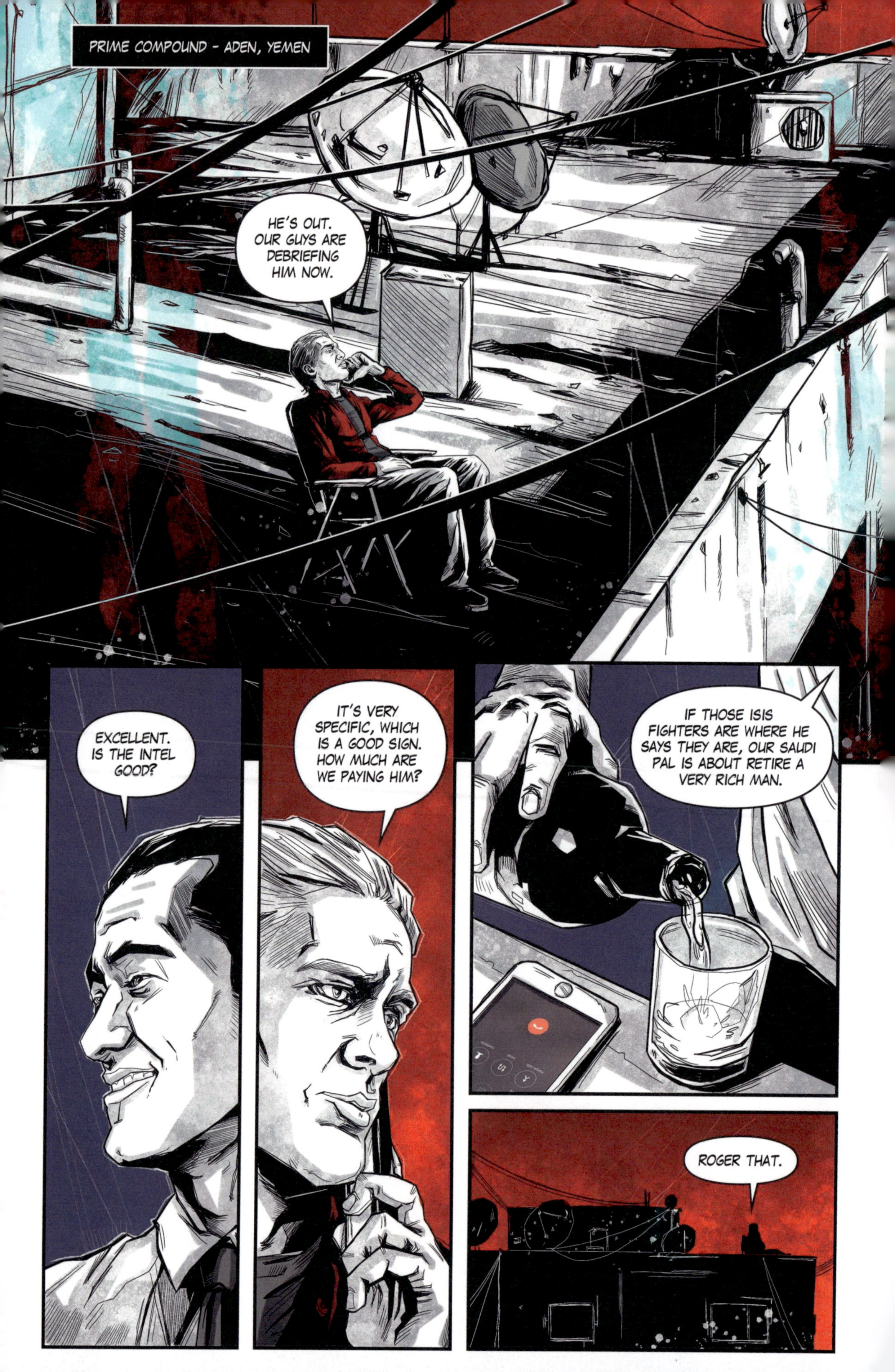
PRIME COMPOUND - ADEN, YEMEN
HE'S OUT. OUR GUYS ARE DEBRIEFING HIM NOW.
EXCELLENT. IS THE INTEL GOOD?
IT'S VERY SPECIFIC, WHICH IS A GOOD SIGN. HOW MUCH ARE WE PAYING HIM?
IF THOSE ISIS FIGHTERS ARE WHERE HE SAYS THEY ARE, OUR SAUDI PAL IS ABOUT RETIRE A VERY RICH MAN.
ROGER THAT.

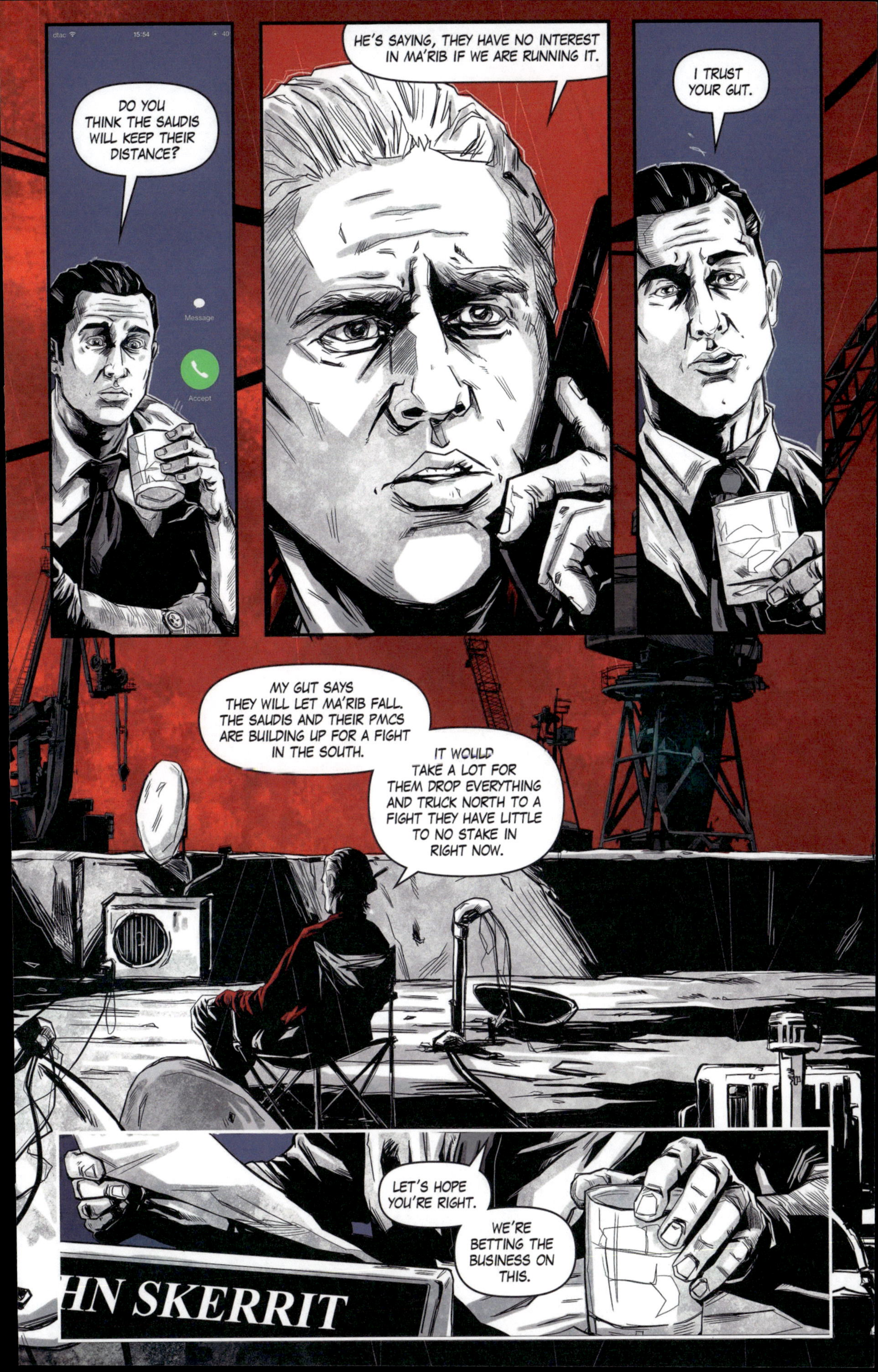
dtac
15:54
DO YOU THINK THE SAUDIS WILL KEEP THEIR DISTANCE?
Message
Accept
HE'S SAYING, THEY HAVE NO INTEREST IN MA'RIB IF WE ARE RUNNING IT.
I TRUST YOUR GUT.
MY GUT SAYS THEY WILL LET MA'RIB FALL. THE SAUDIS AND THEIR PMCS ARE BUILDING UP FOR A FIGHT IN THE SOUTH.
IT WOULD TAKE A LOT FOR THEM DROP EVERYTHING AND TRUCK NORTH TO A FIGHT THEY HAVE LITTLE TO NO STAKE IN RIGHT NOW.
LET'S HOPE YOU'RE RIGHT.
WE'RE BETTING THE BUSINESS ON THIS.
HN SKERRIT

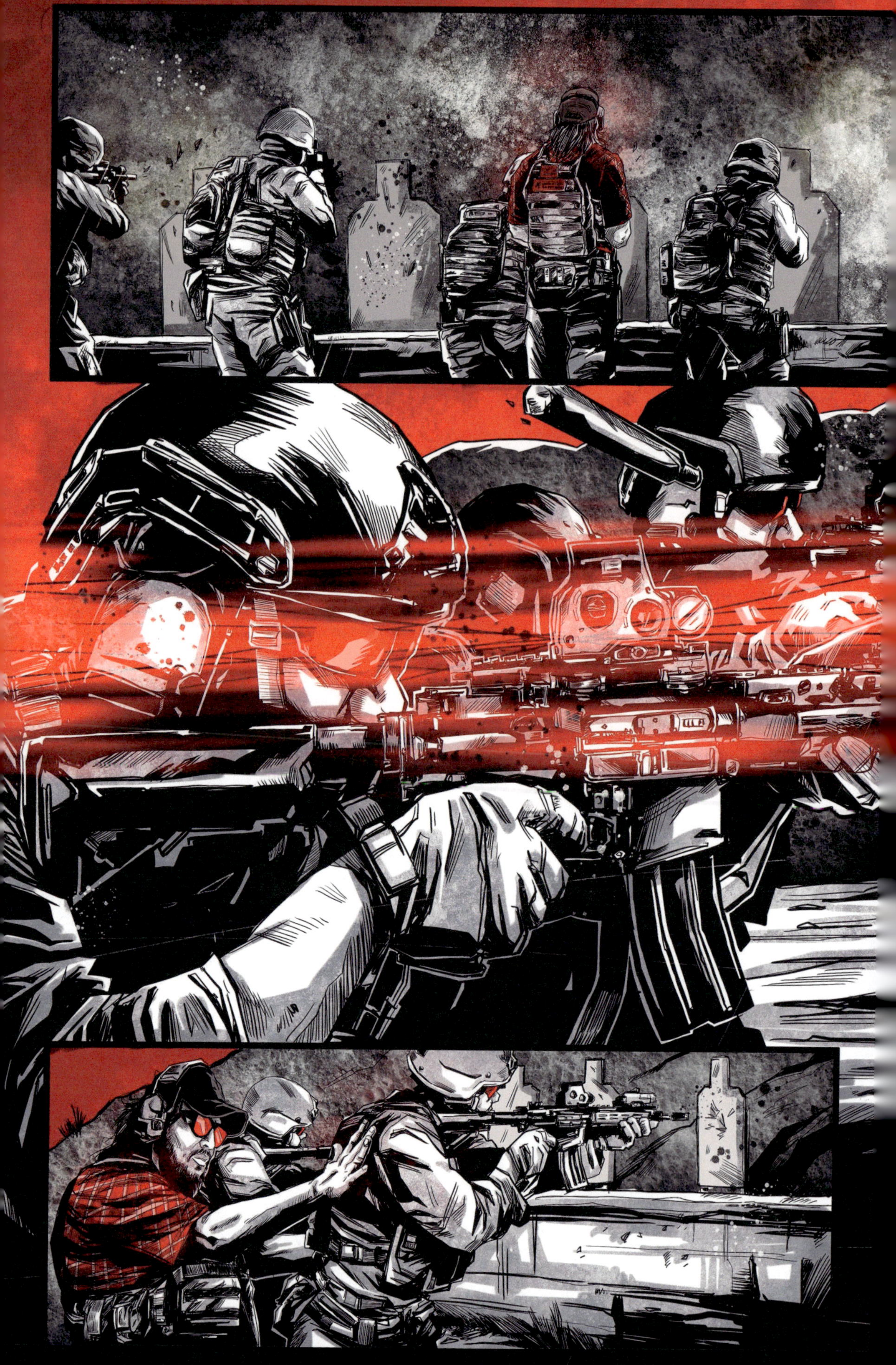

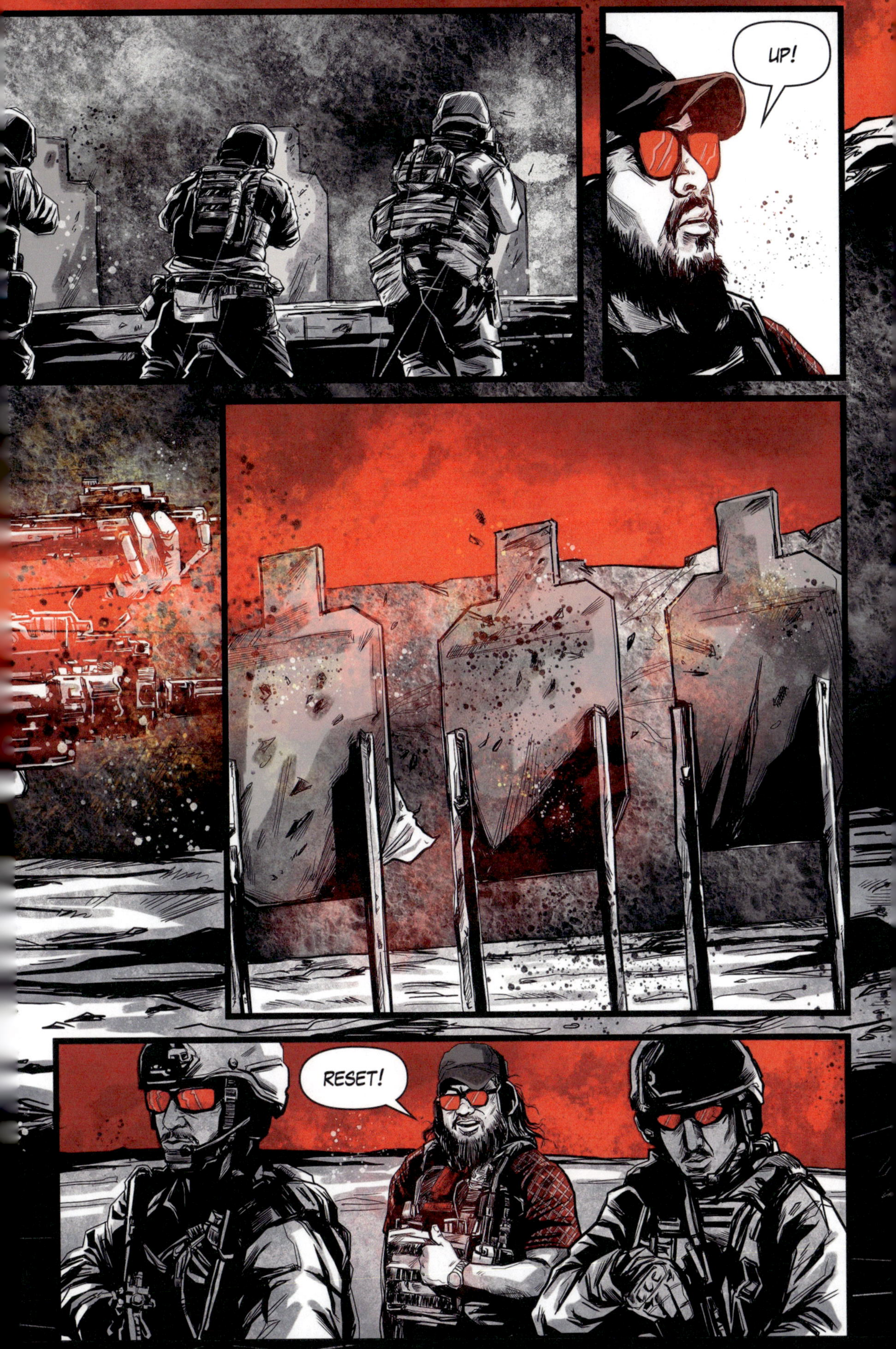
UP!
RESET!

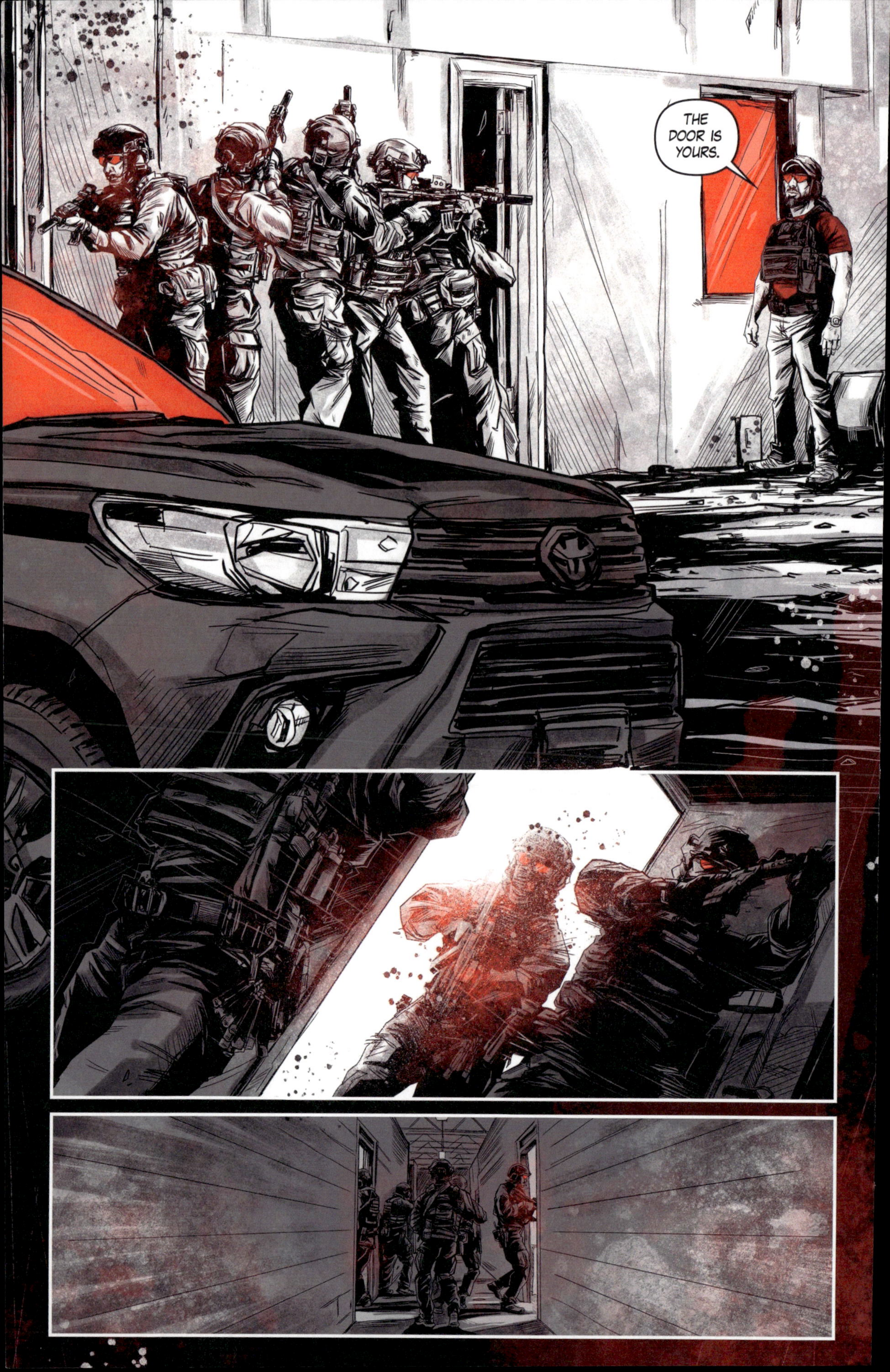
THE DOOR IS YOURS.

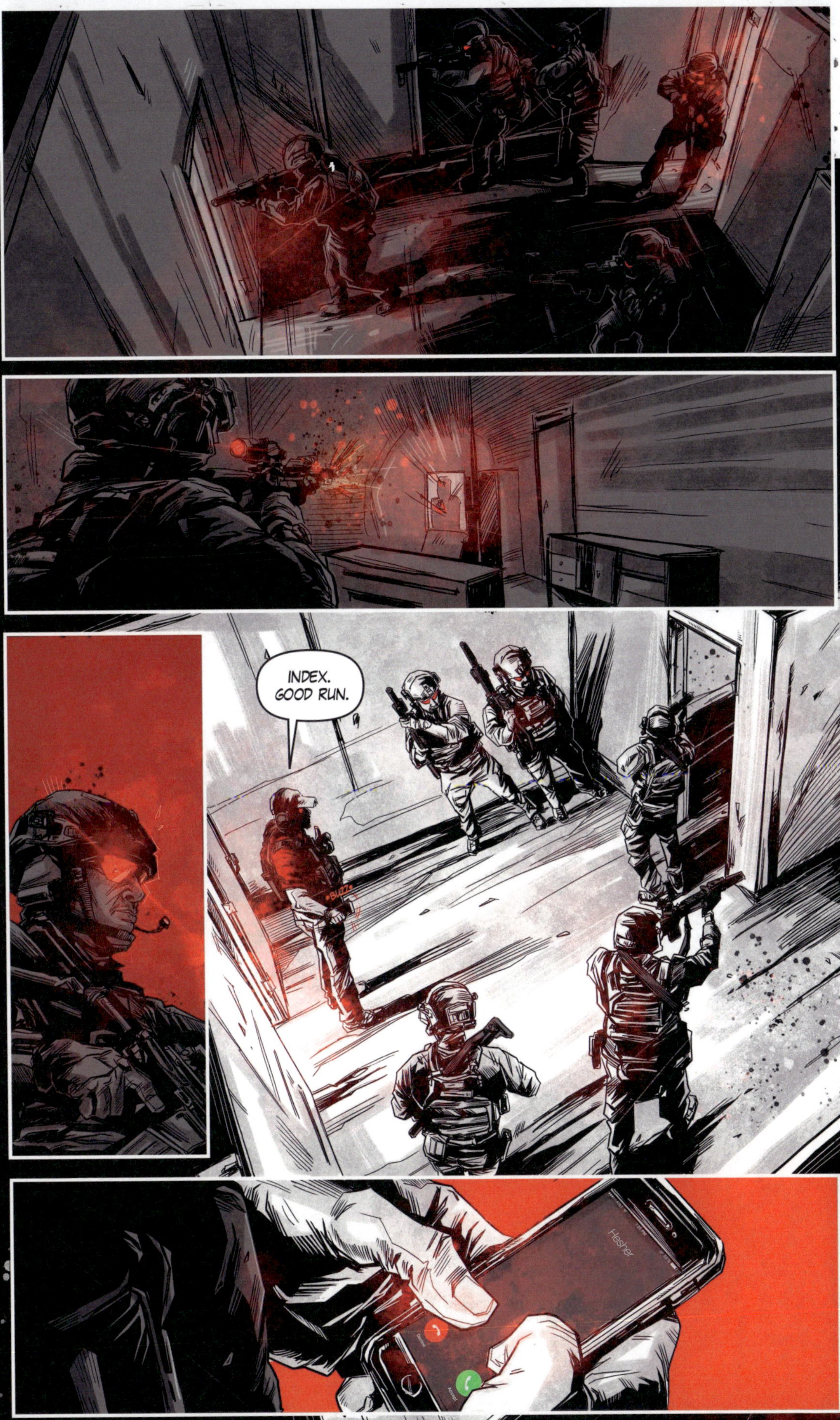
INDEX. GOOD RUN.
•BUZZ•
Hasher

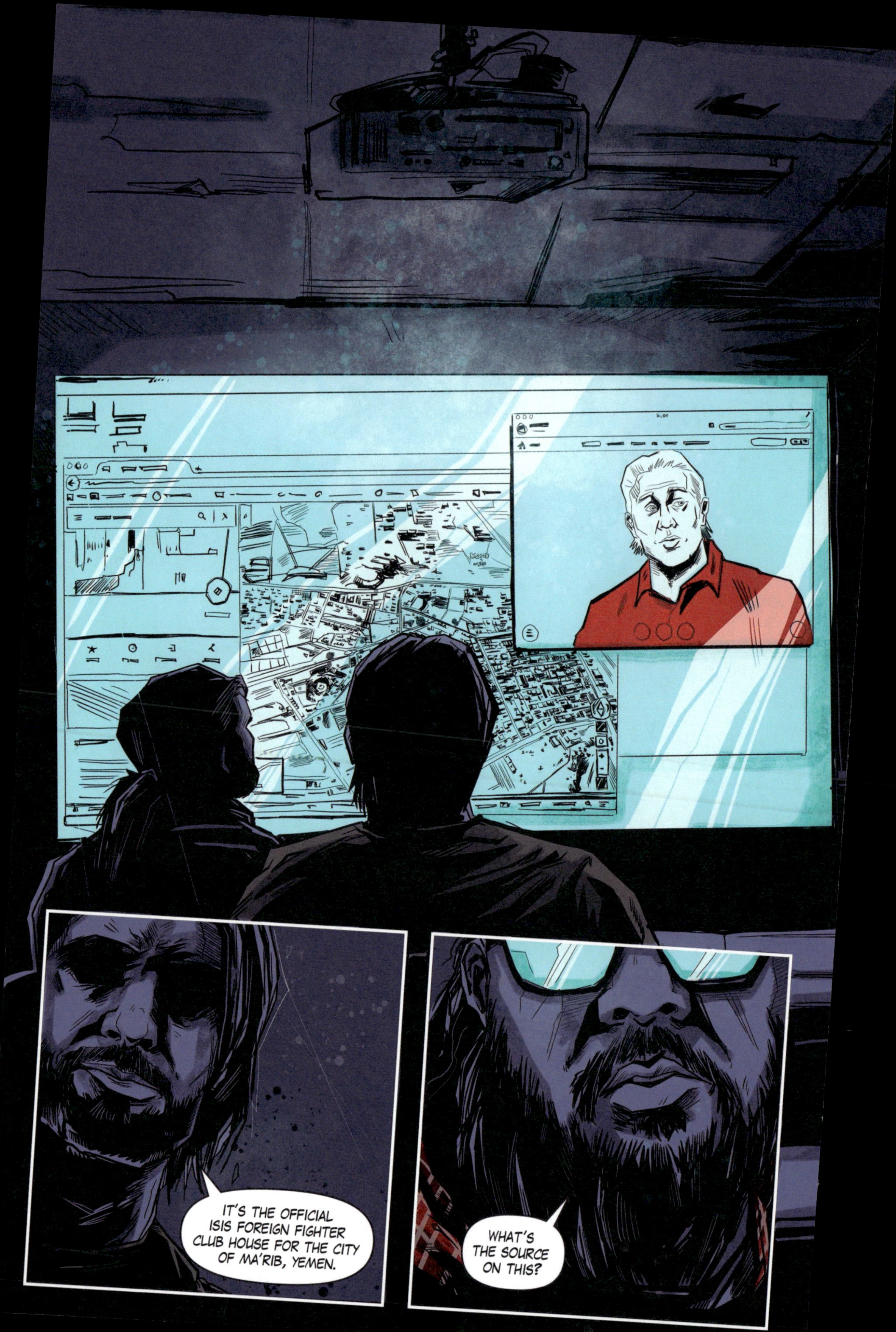
IT'S THE OFFICIAL ISIS FOREIGN FIGHTER CLUB HOUSE FOR THE CITY OF MA'RIB, YEMEN.
WHAT'S THE SOURCE ON THIS?

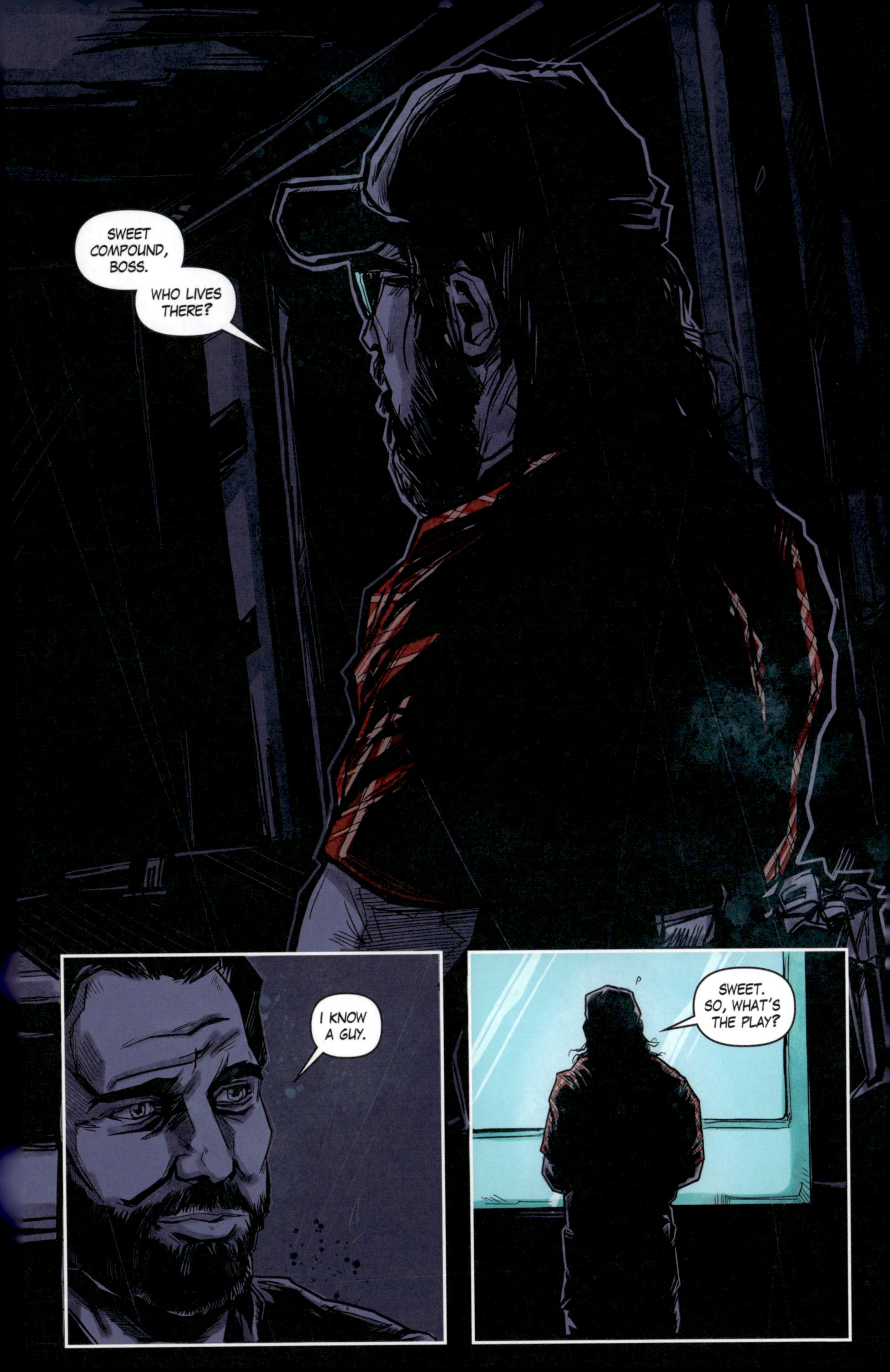
SWEET COMPOUND, BOSS.
WHO LIVES THERE?
I KNOW A GUY.
SWEET. SO, WHAT'S THE PLAY?

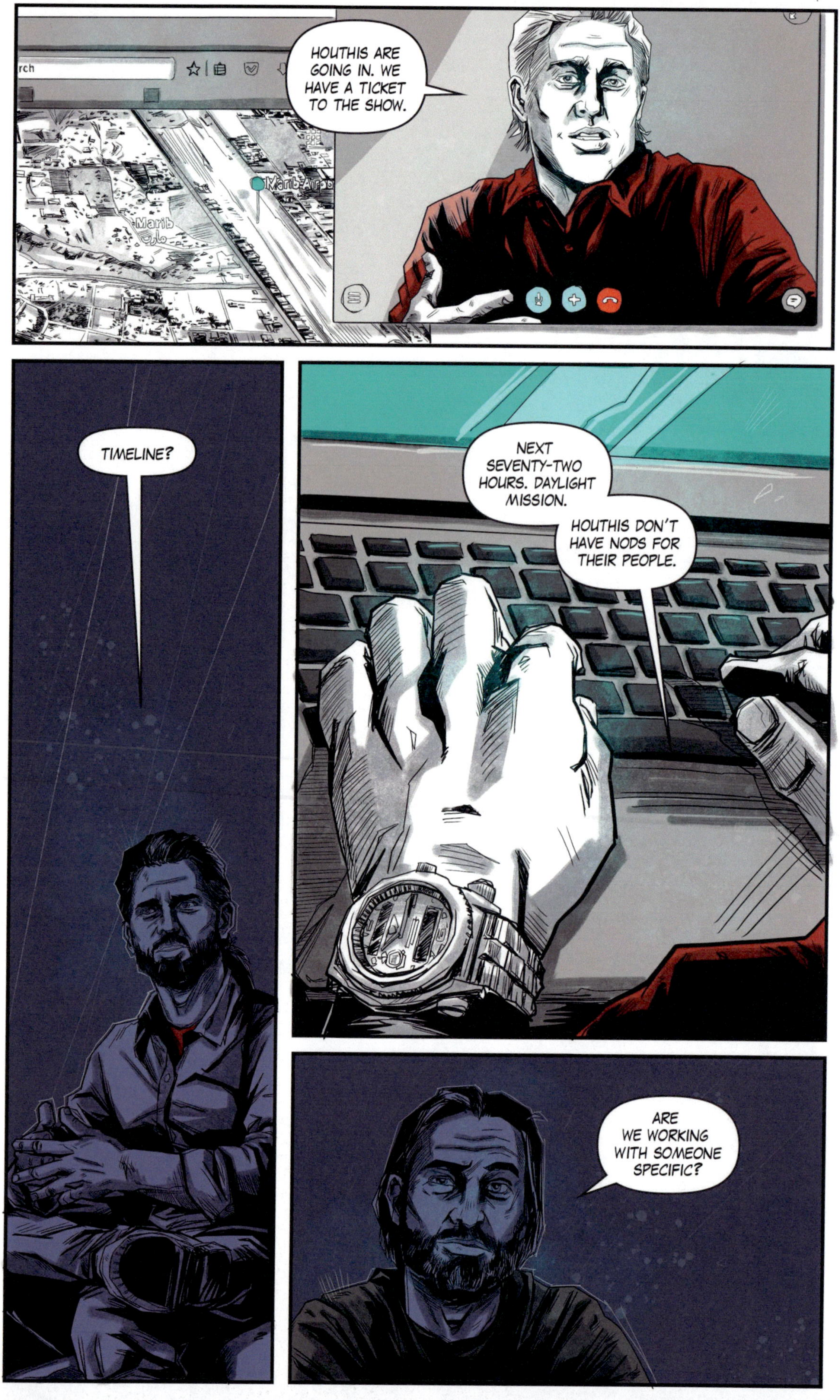
Marib
HOUTHIS ARE GOING IN. WE HAVE A TICKET TO THE SHOW.
TIMELINE?
NEXT SEVENTY-TWO HOURS. DAYLIGHT MISSION.
HOUTHIS DON'T HAVE NODS FOR THEIR PEOPLE.
ARE WE WORKING WITH SOMEONE SPECIFIC?

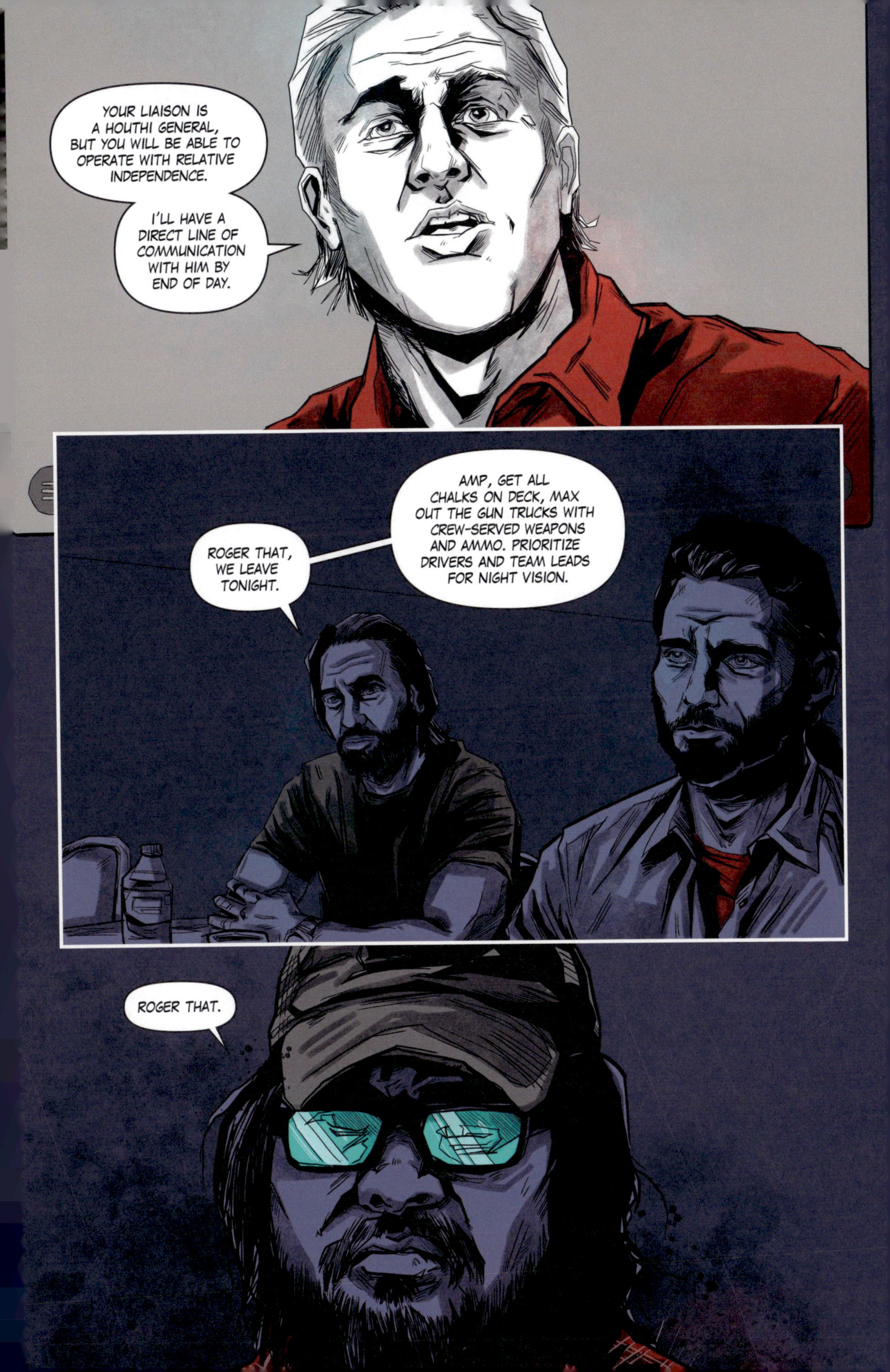
YOUR LIAISON IS A HOUTHI GENERAL, BUT YOU WILL BE ABLE TO OPERATE WITH RELATIVE INDEPENDENCE.
I'LL HAVE A DIRECT LINE OF COMMUNICATION WITH HIM BY END OF DAY.
AMP, GET ALL CHALKS ON DECK, MAX OUT THE GUN TRUCKS WITH CREW-SERVED WEAPONS AND AMMO. PRIORITIZE DRIVERS AND TEAM LEADS FOR NIGHT VISION.
ROGER THAT, WE LEAVE TONIGHT.
ROGER THAT.

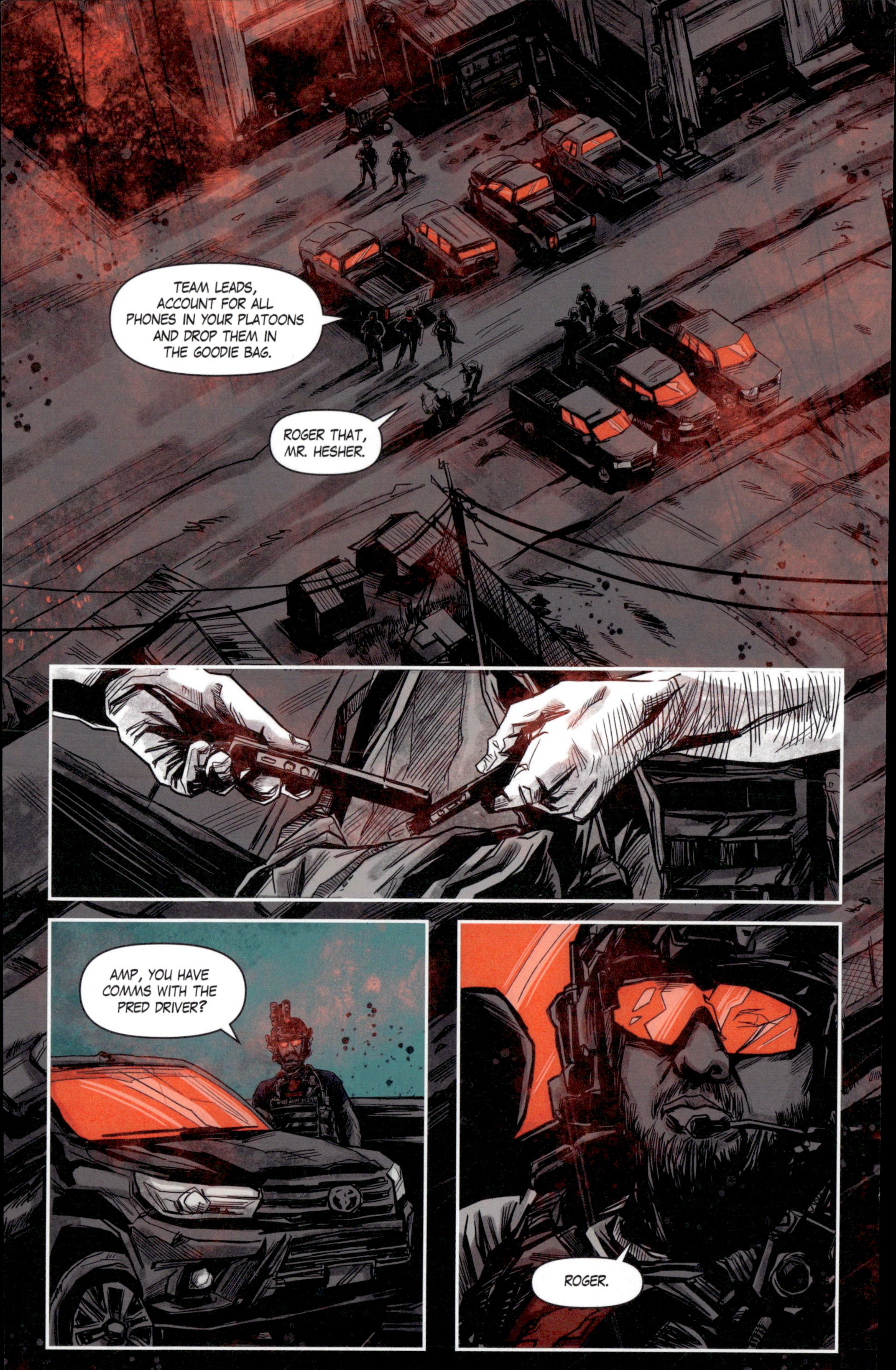
TEAM LEADS, ACCOUNT FOR ALL PHONES IN YOUR PLATOONS AND DROP THEM IN THE GOODIE BAG.
ROGER THAT, MR. HESHER.
AMP, YOU HAVE COMMS WITH THE PRED DRIVER?
ROGER.

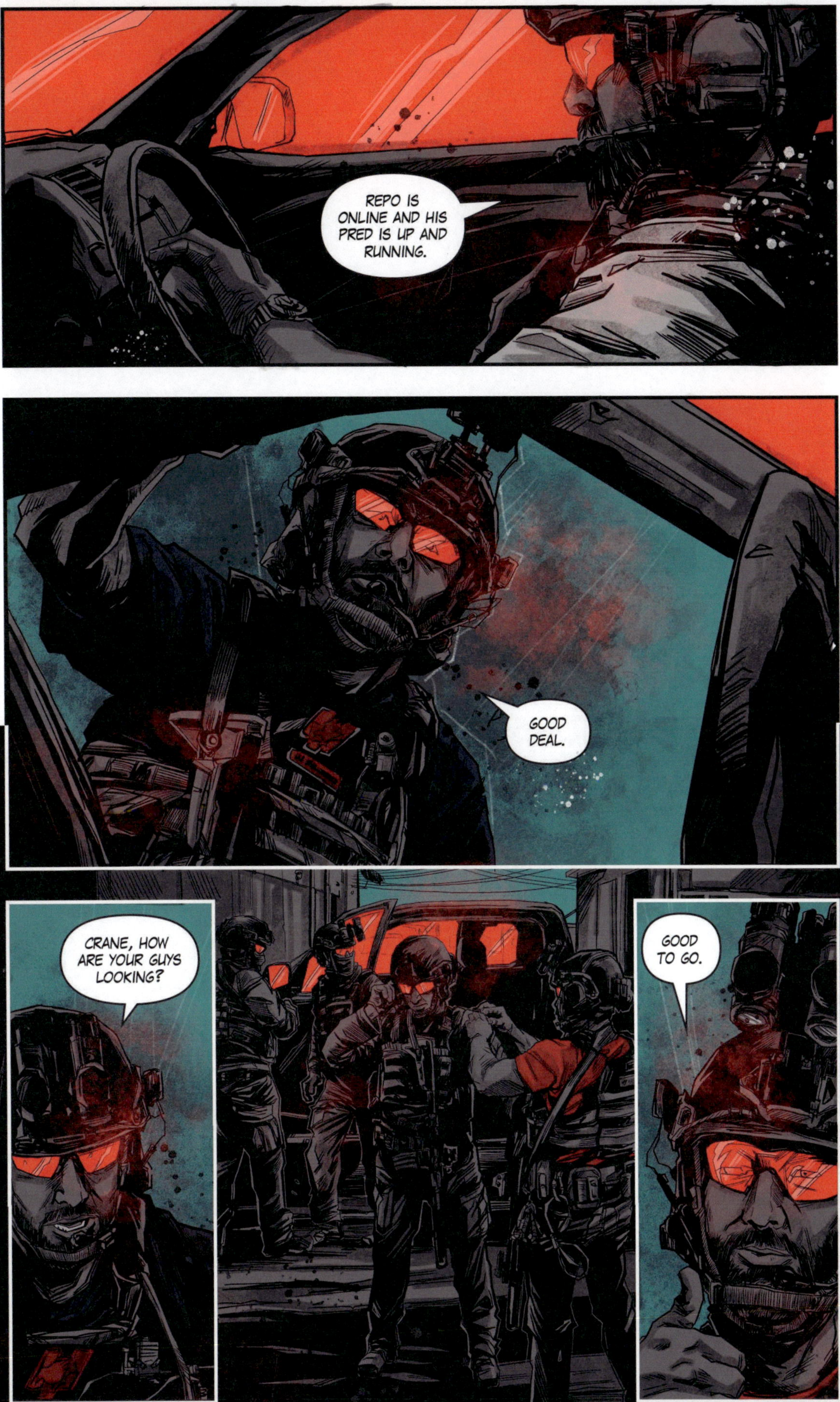
REPO IS ONLINE AND HIS PRED IS UP AND RUNNING.
GOOD DEAL.
CRANE, HOW ARE YOUR GUYS LOOKING?
GOOD TO GO.

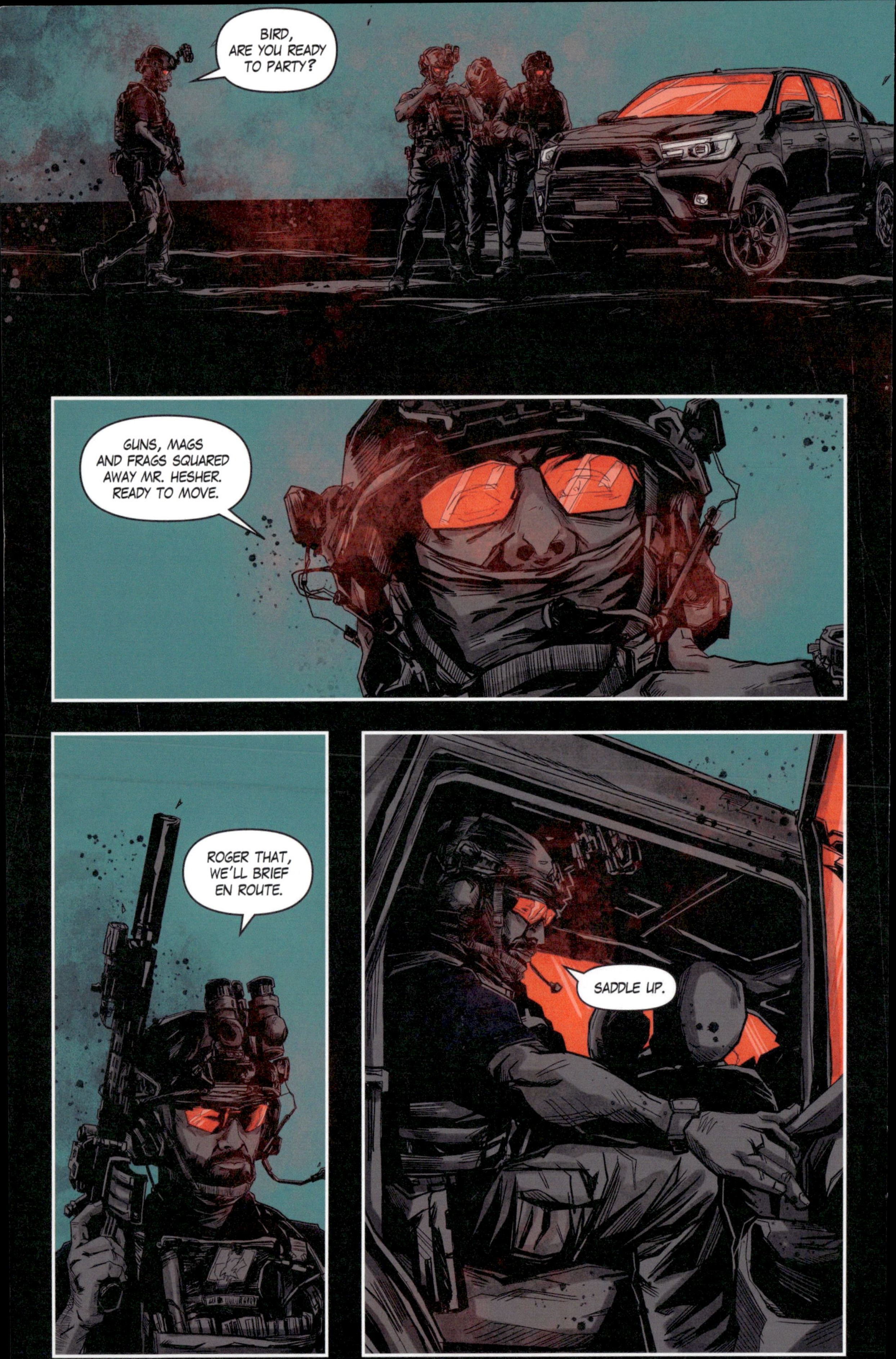
BIRD, ARE YOU READY TO PARTY?
GUNS, MAGS AND FRAGS SQUARED AWAY MR. HESHER. READY TO MOVE.
ROGER THAT, WE'LL BRIEF EN ROUTE.
SADDLE UP.

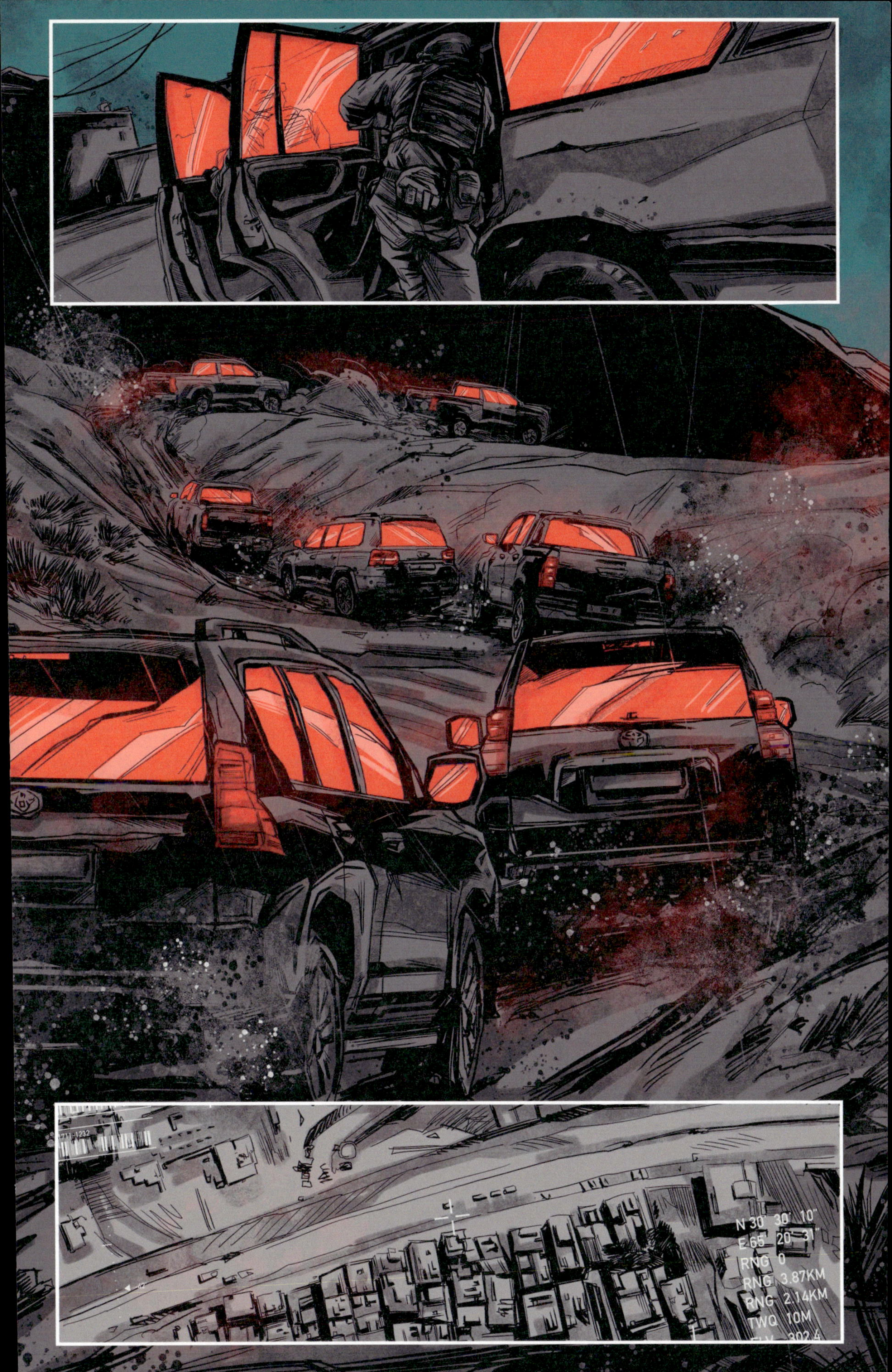
N 30° 30' 10"
E 65° 20' 31"
RNG 0
RNG 3.87KM
RNG 2.14KM
TWQ 10M

SCRIPT
JON CHANG

STORY
JON CHANG

ART & LETTERING
JOSH TAYLOR

LEAD EDITOR
DAVYDD PATTINSON

BLACK POWDER\\RED EARTH LOGO DESIGN BY
JON CHANG

SPECIAL THANKS TO
ECHELON SOFTWARE

COPYRIGHT
©2017 ALL RIGHTS RESERVED

WWW
BLACKPOWDERREDEARTH.COM

SPECIAL THANKS TO OUR PATREON SUPPORTERS
MIKE MEDNANSKY
THOMAS COMPITELLO
SAMUEL MICHLOVSKY
RUBEN GONZALO
MITSUHISA TAKEMORI
SEAN MARIUCCI
LEE TOLLESON

Made in the USA
Columbia, SC
18 March 2024